BUILDING ORGANIZATIONAL INTELLIGENCE
A Knowledge Management Primer

BUILDING ORGANIZATIONAL INTELLIGENCE
A Knowledge Management Primer

Jay Liebowitz
Robert W. Deutsch, Distinguished Professor of Information Systems
University of Maryland, Baltimore County
Department of Information Systems
Baltimore, Maryland

CRC Press
Taylor & Francis Group
Boca Raton London New York

CRC Press is an imprint of the
Taylor & Francis Group, an **informa** business

CRC Press
Taylor & Francis Group
6000 Broken Sound Parkway NW, Suite 300
Boca Raton, FL 33487-2742

© 2000 by Taylor & Francis Group, LLC
CRC Press is an imprint of Taylor & Francis Group, an Informa business

First issued in paperback 2019

No claim to original U.S. Government works

ISBN-13: 978-0-367-45562-0 (pbk)
ISBN-13: 978-0-8493-2036-1 (hbk)

This book contains information obtained from authentic and highly regarded sources. Reasonable efforts have been made to publish reliable data and information, but the author and publisher cannot assume responsibility for the validity of all materials or the consequences of their use. The authors and publishers have attempted to trace the copyright holders of all material reproduced in this publication and apologize to copyright holders if permission to publish in this form has not been obtained. If any copyright material has not been acknowledged please write and let us know so we may rectify in any future reprint.

Except as permitted under U.S. Copyright Law, no part of this book may be reprinted, reproduced, transmitted, or utilized in any form by any electronic, mechanical, or other means, now known or hereafter invented, including photocopying, microfilming, and recording, or in any information storage or retrieval system, without written permission from the publishers.

For permission to photocopy or use material electronically from this work, please access www.copyright.com (http://www.copyright.com/) or contact the Copyright Clearance Center, Inc. (CCC), 222 Rosewood Drive, Danvers, MA 01923, 978-750-8400. CCC is a not-for-profit organiza-tion that provides licenses and registration for a variety of users. For organizations that have been granted a photocopy license by the CCC, a separate system of payment has been arranged.

Trademark Notice: Product or corporate names may be trademarks or registered trademarks, and are used only for identification and explanation without intent to infringe.

Visit the Taylor & Francis Web site at
http://www.taylorandfrancis.com

and the CRC Press Web site at
http://www.crcpress.com

Cover design: Dawn Boyd

Library of Congress Card Number 99-29263

Library of Congress Cataloging-in-Publication Data

Liebowitz, Jay
    Building organizational intelligence : a knowledge management primer / Jay Liebowitz
        p.   cm.
    Includes bibliographical references and index.
    ISBN 0-8493-2036-4 (alk. paper)
    1. Organizational learning. 2. Knowledge management. I. Title.
HD58.82.L53 1999
658.4'038—dc21

99-29263
CIP

Dedication

Dedicated to Janet, Jason, Kenny, and knowledge sharers worldwide

Table of Contents

1. What is Organizational Intelligence? 1
2. Creating Knowledge 11
3. Capturing and Storing Knowledge 19
4. Transforming Individual Learning into Organizational Learning 31
5. Combining, Transferring, and Distributing Knowledge 37
6. Building a Continuous Learning Culture 41
7. Culture: The Key Ingredient 49
8. Developing a Knowledge Management Capability 55
9. Assessing Knowledge Management Through a Knowledge Audit 63
10. Augmenting Organizational Intelligence 67

Role and Skills for Knowledge Management — Questionnaire 73

Appendix A: The Intelligent Agent-Based Knowledge Management System for Supporting Multimedia Systems Design on the Web 85

Appendix B: Audit of Knowledge Management Practices for Innovation 113

Appendix C: Knowledge Management: Strategies for the Learning Organization Screening Survey 121

Appendix D: Valuating Knowledge Assets (Human Capital) Knowledge Survey 135

Index 139

The Author

Dr. Jay Liebowitz is the Robert W. Deutsch Distinguished Professor of Information Systems at the University of Maryland — Baltimore County, and formerly served as Professor of Management Science in the School of Business and Public Management at George Washington University in Washington, D.C. He recently served as Chaired Professor in Artificial Intelligence at the U.S. Army War College. He was selected as the 1996 Computer Educator of the Year by the International Association for Computer Information Systems. He is the Editor-in-Chief of the international journals *Expert Systems with Applications* (Elsevier) and *Failure & Lessons Learned in Information Technology Management* (Cognizant Communication Corporation). Liebowitz is the founder and Chair of the World Congress on Expert Systems, and a Fulbright Scholar and recipient of the IEEE-USA FCC Executive Fellowship.

Dr. Liebowitz can be reached at liebowit@umbc.edu.

1 What is Organizational Intelligence?

Most Chief Executive Officers feel that knowledge is the most critical asset of their organization. They believe that their organization's "brainware", the intellectual capital of their firm, is what gives the organization their competitive edge. As one CEO from Texas Instruments said, "If we only knew what we know!"

In today's movement towards knowledge management, organizations are trying to best leverage their knowledge internally in the organization and externally to their customers and stakeholders. They are trying to capitalize on their "organizational intelligence" to maintain their competitive edge.

Companies are realizing that their human capital and structural capital are the distinguishing elements of their organization. Human capital refers to the "people power" and structural capital is databases, patents, intellectual property, and related items that the employee can't readily take home with them.

Knowledge management is one of the "hottest" terms in organizations today. The thrust of knowledge management is to create a process of valuing the organization's intangible assets in order to best leverage knowledge internally and externally. Knowledge management, therefore, deals with creating, securing, capturing, coordinating, combining, retrieving, and distributing knowledge. The idea is to create a knowledge sharing environment whereby "sharing knowledge is power" as opposed to the old adage that, simply, "knowledge is power."

In order to create knowledge management systems, it is critical to build a supportive culture (from top management on down). This was certainly true in the Buckman Labs example of their K'Netix knowledge management

An organization's competitive edge

Employee/management
brainpower/knowledge
and overall intellectual
capital of the organization

system, wherein Bob Buckman (CEO of Buckman Labs) said that 90% of knowledge management and their success was building the culture to encourage knowledge sharing. In a recent benchmarking study of about 150 companies, the reason given by those individuals who didn't want to share their knowledge was not that they wanted to keep their competitive edge close to heart. Rather, it was that they wouldn't be able to put their own thumbprint on knowledge if they had to use someone else's knowledge. This suggests the need for an incentive and reward system to encourage knowledge sharing. Some companies, like Andersen Consulting and Lotus, evaluate their employees for their annual job performance review partly on how well they contribute their knowledge to the organization's knowledge repositories and how well they use and apply the knowledge that exists in these repositories. Oftentimes, the technology is not the limiting factor in creating a knowledge management environment — it's the "management" of the technology and "culture" that greatly influences the success of knowledge management endeavors.

The intelligent systems discipline can be applied easily to the knowledge management field. Case-based reasoning, for example, is an effective technology for help-desk applications. Tools exist, like CasePoint by Inference, to facilitate the development of these case-based help desks. Expert systems can also be used to help capture the experience and knowledge of those experts within the organization before they retire or leave. The ELAWS effort in the U.S. Department of Labor is an excellent step in this direction in developing knowledge compliance systems over the web as employment law advisors for workers and small businesses. Data mining may also be an appropriate technology to help determine trends and relationships in data and information. The use of intelligent agents can also be applied for helping better tailor the search and navigation process of important information and documents. Additionally, all these technologies should be designed to conform with proper human–computer interaction guidelines with respect to cognitive overloads, visualization, and user interface considerations.

Modest knowledge management efforts could commence in organizations with a "yellow page" directory of mapping knowledge areas to experts within the organization. A best-practices or lessons-learned knowledge repository could also be created to facilitate knowledge sharing efforts. There should be a knowledge management infrastructure within the organization whose mandate is to identify, analyze, manage, maintain, and disseminate knowledge to appropriate individuals within the organization and externally to others. Many organizations already have created a "Chief Knowledge Officer" position or equivalent to help spearhead these knowledge management efforts. Through web-based and intranet technologies, we now have the "connectivity" to allow the "collection" of knowledge bases to be disseminated more easily than before. This will help contribute towards building an entity's "organizational intelligence" (OI).

Knowledge Management: Is It So New?

Knowledge management (KM) is the process of creating value from an organization's intangible assets. Is this concept really new? Not really! It's an amalgamation of knowledge-based systems, artificial intelligence, software engineering, business process improvement, human resources management, and organizational behavior concepts.

So why the craze? Companies are realizing that their competitive edge is mostly the brainpower or intellectual capital of their employees and management. Many organizations are drowning in information and starving for knowledge. In order to stay ahead of the pack, companies must leverage their knowledge to survive. With web-based and intranet technologies, the connectivity and possible sharing of knowledge are greatly enabled in order to build the knowledge infrastructure of the firm. Knowledge management is believed to be the current savior of organizations, but it is much more than developing Lotus Notes lessons-learned databases. Knowledge management deals with the conceptualization, review, consolidation, and action phases of creating, securing, combining, coordinating, and retrieving knowledge.

So what makes knowledge management so hard? First, the organization must create a knowledge sharing environment. Some firms provide incentives to promote this climate until it becomes the norm. Other organizations require their employees to contribute actively and use knowledge in the organization's knowledge repositories as part of their annual job performance review.

The second enigma is determining how senior executives can value the knowledge in their organization in order to show some tangible benefits. A number of individuals like Leif Edvinsson, Michael Malone, Karl Sveiby, Tom Stewart, Annie Brooking, Rob van der Spek, and Robert de Hoog, and others have developed methodologies to value knowledge. Some of these techniques value knowledge at the "global" (firm-wide) level, and others value knowledge at the "knowledge item" (lower level). It's not easy to value the intellectual capital (especially the "human capital") in the organization. Unless we develop ways to do this, top management may not place much emphasis on knowledge management, and may not stress the importance of intellectual capital in the organization.

A third fly in the ointment is the belief that information management is the same as knowledge management. Knowledge is information with a process applied to it, which may eventually become wisdom or expertise. Many organizations are having their IT (information technology) directors become Chief Knowledge Officers, because top management often feels they are comparable positions. This is a mistake because knowledge management draws from many disciplines, including IT, and it is broader in scope than the technology functions that an IT director often oversees. A new breed of knowledge officers or knowledge analysts is needed to fill the roles of knowledge managers in organizations.

A fourth fallacy that organizations haven't fully realized is knowledge management works best when the CEO on down actively use the knowledge management systems designed for their organizations. Without senior management commitment and involvement, the knowledge management systems and infrastructure may be pushed aside and not be integrated within the mainstream of the organization. Buckman Labs' K'Netix (its knowledge management network) was successful largely due to the backing and usage of it by its CEO, Bob Buckman.

The last major concern regarding the survivability of knowledge management is the misnomers being labeled on almost every tool as a "knowledge management" tool. This hype will kill the "good" from knowledge management principles in the same way that the previous fad, BPR (Business Process Reengineering), died out. Many consulting firms are proclaiming their expertise in knowledge management. But the truth of the matter is that knowledge management, as a field, is almost too young for there to be many experts.

Without organizations fully understanding these five major concerns, the fear is that the mystique of knowledge management will remain cloudy and shapeless. With more researchers and practitioners working together to further define and develop the KM field, the mystique of KM will begin to produce an aura of fundamentally sound principles, concepts, methodologies, techniques, and tools. By working together, knowledge management can help build the "organizational intelligence" (OI) of a firm.

Organizational Intelligence

Various definitions have been applied to "organizational intelligence". Some of these definitions include:

- the problem of gathering, processing, interpreting, and communicating the technical and political information needed in the decision-making process (Wilensky);
- the organization's ability to deal with complexity, that is, its ability to capture, share, and extract meaning from marketplace signals (Haeckel and Nolan);
- that capacity for computation which can be applied to information that is externally gained or internally generated to meet survival challenges (McMaster);
- information processing functions that permit adaptation to environmental demands and are related to innovation initiation and implementation (Glynn);
- the intelligent behavior of organizations as a function of their design (Nonaka);
- the cognitive functioning through which information flows through organizations (Halal and Kull);
- understanding organizations as learning systems and creative systems (Nevis; Mumford and Gustafson);

6   Building Organizational Intelligence: A Knowledge Management Primer

- organizational intelligence is a function of five cognitive subsystems: organizational structure, culture, stakeholder relationships, knowledge management, and strategic processes (Halal et al.).

The author's view is that organizational intelligence is the collective assemblage of all intelligences that contribute towards building a shared vision, renewal process, and direction for the entity. Specifically, organizational intelligence involves the following knowledge functions:

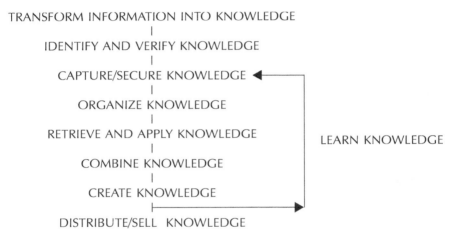

Transforming information into knowledge involves the synthesis and conversion of useful data and information into knowledge. For example, rules of thumb acquired over years of experience and learning may result in knowledgeable shortcuts to help in the decision-making process. Other types of knowledge (i.e., procedural, declarative, episodic, metaknowledge) need to be identified in the organization and verified as relevant knowledge. Once knowledge is identified, it should be captured or acquired and then secured within the organization. Once captured and secured, it must be organized in a way in which others in the organization can retrieve this knowledge and apply it to their situations. They will also combine this knowledge with knowledge of their own within the context of their situation. This will, hopefully, result in a synergistic way of creating new knowledge for the organization. This knowledge would then be distributed within the organization or to stakeholders and possibly sold. New knowledge would, hopefully, be learned, captured, and secured within the organization, and then the cycle would continue.

Organizational intelligence would be built, adapted, and refined from this cyclical process. Of course, a series of questions result in order to build and leverage an organization's intellectual capital, as developed by Linkage Inc.:

- What is your organization doing to build systems that create, capture, and leverage customer knowledge into business opportunities?
- What systems are in place to capture the tacit knowledge and expertise of employees who are leaving your organization?
- Do your organization's reward and/or recognition practices motivate people to collaborate and share knowledge?
- How much work is duplicated because knowledge, information, and lessons learned are not accessible to all?
- Is your training and learning strategy aligned with the needs of a knowledge enterprise?
- Is there an opportunity for you to lead your organization's knowledge and learning strategy?
- How do you assess your organization's learning culture?

An important part of organizational intelligence is the renewal process in fostering organizational learning within the entity. The ability to transform individual learning into organizational learning is a challenge in the organization. Several new organizational forms and metaphors are being used to assist in the learning and knowledge deployment process. One useful technique is through a "Community of Practice". John Seely Brown, Director of Xerox PARC in 1993, defined "Community of Practice" as:

> At the simplest level, they are a small group of people who've worked together over a period of time...not a team, not a task force, not necessarily an authorized or identified group...perform the same tasks...or collaborate on a shared task...or work together on a product...They are peers in the execution of "real work". What holds them together is a common sense of purpose and a real need to know what each other knows.

According to the Institute for Research on Learning, a spin-off of Xerox Corporation, a Community of Practice has the following characteristics: (1) learning is fundamentally a social phenomenon; (2) knowledge is integrated in the life of communities that share values, beliefs, languages, and ways of doing things; (3) the process of learning and the process of membership in a community of practice are inseparable; (4) knowledge is inseparable from practice; and (5) the ability to contribute to a community creates the potential for learning. Intel, Dow Chemical, and National Semiconductor support Communities of Practice.

The Role of Organizational Learning in Organizational Intelligence

Organizational learning is a key component of organizational intelligence (OI). In order for an organization's intelligence to increase, the organizational entity must be able to learn and adapt. Even by having the ability to learn and adapt, this may not guarantee that the organization's intelligence will grow. It is highly likely that the OI will flourish, but some inhibiting factors may exist. These might include: a major undetected competitor entering the marketplace; an unusual economic downturn; poor strategic planning; or perhaps poor management acumen and personnel. The ability of the organization to have sensors and systems in place to be able to predict, anticipate, react, and adjust to changes is an aspect of both organizational learning and OI.

At the 1998 Conference on Knowledge Management and Organizational Learning, sponsored by IBM and The Conference Board, several important ideas were highlighted:

- The only way to sustain competitive advantage is to ensure that your organization is learning faster than the competition.
- Whereas knowledge should be seen as a resource and asset, learning should be seen as the process which really drives us. Learning should be our focus and the main strategic requirement.
- Knowledge is socially constructed within collaborative activity.
- Access does not equal success.

A key concept emanating from this conference, and expressed by Larry Prusak of IBM Global Services, is knowledge originates in people and becomes "embodied" in networks and communities, "embedded" in work routines, practices, and norms, and "represented" in artifacts (e.g., documents, products, files, reports, etc.). In order to build knowledge capital and thus organizational intelligence, social capital is probably the most critical element. As Prusak explains, social capital involves:

- Trust ("anticipated reciprocity" or "residue of promises kept")
- Space (cognitive space — share of mind around knowledge)
- Slack (time to reflect on what you know)
- Coherence (shared context, shared language, "communities of
- practice")
- Shared vocabulary
- Symbols and signals

Social capital contributes to organizational learning. Essential characteristics of organizational learning include: new skills, attitudes, values, behaviors, and products created or acquired over time; what is learned becomes the property of some collective unit; what is learned remains within the organization or group even if individuals leave.

According to DiBella and Nevis in their 1998 book titled *How Organizations Learn: An Integrated Strategy for Building Learning Capability*, they feel that organizational learning and expertise are comprised of the following activities: collaborate across functions, encourage exploration and experimentation, identify best practices, create and maintain knowledge repositories, create centers of expertise, educate/consult/coach workforce, and develop performance support systems.

Building a sharing culture is a necessary element for organizational learning and organizational intelligence. This may mean transforming a verbal to a written sharing culture, and providing ways to encourage this culture to exist. At American Management Systems (AMS), individual success is measured in terms of leverage. There are awards for the most reused item in a year and there are knowledge-in-action rewards for creative contribution and reuse of an asset. At AMS, Corporate Knowledge Centers exist in core competency areas of AMS. AMS has learned the following lessons: recognize individual achievement, build group identity, motivate and reward performance, celebrate successes, and deliver value.

According to the American Productivity and Quality Center report titled "If We Only Knew What We Know: Identification and Transfer of Internal Best Practices", there are several obstacles that make organizational learning difficult. These include:

- Organizational structures that promote "silo" thinking, in which locations, divisions, and functions focus on maximizing their own accomplishments and rewards, hoarding information, and thereby suboptimizing the total organization

- A culture that values personal technical expertise and knowledge creation over knowledge sharing
- The lack of contact, relationships, and common perspectives among people who don't work side-by-side
- An over-reliance on transmitting "explicit" rather than "tacit" information
- Not allowing or rewarding people for taking the time to learn and share and help each other outside of their own small corporate village

If we can recognize and address these stumbling blocks, organizations will be able to create a more fertile knowledge-sharing environment, leading to (hopefully) increased organizational learning and a higher organizational intelligence.

In the chapters ahead, we will address the various components dealing with organizational intelligence. Here we go...

2 Creating Knowledge

In order to develop organizational intelligence (OI), there must be processes and systems in place to facilitate the "creation" of knowledge. Glynn (1996) provides a definition for organizational intelligence:

Organizational intelligence is an organization's capability to process, interpret, encode, manipulate, and access information in a purposeful, goal-directed manner, so it can increase its adaptive potential in the environment in which it operates.

Glynn further describes organizational intelligence as having elements of: adaptive, purposeful information processing; context-specific dominant intelligence; social outcome. She describes various models and measures that have been used to assess organizational intelligence. These include:

- Aggregation Model: Intelligence of individual organizational members aggregates as organizational intelligence.
  —Measurements: Organizational intelligence may be assessed as the aggregated total, the average, or the maximum of individual intelligence.
- Cross-Level Model: Individual intelligence is transferred and encoded in organizational systems to become organizational intelligence.
  —Measurements: Effectiveness of mechanisms that transfer and institutionalize intelligence, including structural, technical, social and/or political influence mechanisms.
- Distributed Model: Organizational intelligence is embedded in the organization's systems, routines, standard operating procedures, symbols, culture, and language.
  —Measurements: Richness and ecological validity of an organization's systemic interaction patterns; assessment of reliability of

systems; assessment of an organization's behavior interactions through qualitative studies.

In a recent study of Fortune 100 firms conducted by the Strategic Decisions Group in New York, the level of "value identified" was about 50 percent and "value delivered" was only 25 percent. This equates to organizations achieving only 12.5 cents on the dollar in terms of value-creation. As stated in the 1998 New York Business Information Conference, 90 percent of revenues from an organization typically come from the organization's "intangible" assets. The intangible assets would refer to human capital, structural capital, and the customer capital of an organization. Additionally, Larry Prusak of IBM indicates that 70–80 percent of learning is through "informal" means vs. formal artifacts (documents, reports, books, etc.).

These numbers indicate that an organization needs to better tap into the human and related capital of its firm to achieve a higher level of value creation and to improve upon ways of creating, capturing, combining, distributing, and leveraging this knowledge. One of the main bottlenecks of developing new knowledge, according to Rob van der Spek of CIBIT in Utrecht, is the following:

> Not enough is learned from developments in the market. Knowledge about potential markets, current markets and present or new competitors is not structurally developed. The structured development of new ideas into products that are ready for the market does not take place. This is often because there is insufficient patience and commitment to give new ideas a chance. In many cases, there is also no clear vision of the future, so that it is actually impossible to determine which ideas need to be worked out.

American Management Systems, a large consulting firm based in Fairfax, Virginia, has been involved with knowledge management since 1993. Two of their knowledge management initiatives include the AMS Best Practices Program and the AMS Knowledge Centers (knowledge-based communities of practices). The Best Practices' consulting activities help AMS harvest new knowledge. There are also Associates (expert practitioners) of AMS Knowledge Centers who serve as communities of experts in AMS' core disciplines (namely, System Development and IT Management, Business Process Renewal, Organization Development and Change Management, Advanced Technologies, Engagement and Project Management, Customer Value Management, Knowledge Management, and Electronic Commerce). To encourage the creation and sharing of knowledge, AMS has a motivate-and-reward system whereby individual success is measured in terms of leverage, and a group identity is built via these sharing cultures. AMS celebrates successes via Best Practices Awards for creation and reuse of knowledge assets, media recognition, contests, awards, annual conferences for the community members, reuse rate for contributions publicized, and other mechanisms.

Ray Manganelli, Director of Strategic Decisions Group in New York, indicated five key elements to the 21st century company. These included: developing a value-creation culture, applying a decision quality framework, focusing on distribution channels, transitioning from "make and sell" to "sense and respond", and aligning the value delivery platform. In order to ready organizations for the new millenium, organizations must be adaptable and have strong connectivity and partnership with their customers. Through these interactions, knowledge should be created in order to continue growth of the organization and its associated organizational intelligence.

Hubert St. Onge, Senior Vice President for Strategies at The Mutual Group in Canada and a leader in the intellectual capital field, states the following key propositions:

- The leadership and culture of a firm determine its capability to develop relationships with customers
- An organization's ability to build customer relationships depends on the readiness of its leadership to engage in the development of the required organizational culture
- Values measurement can inform us about the readiness of leadership to make the necessary shifts
- Values reside at the core of an organization's tacit knowledge: making these values explicit builds an effective platform for accelerating capability building
- The place where leadership and culture come together is in the necessity of the enterprise to commit to a clear vision that aligns leadership, culture, and branding

Sandra Ward, Director of Information Services at Glaxo Wellcome R&D in the United Kingdom, echoes some of the St. Onge's propositions and indicates that knowledge sharing competencies include the creation, capture, communication, and use of knowledge. The "create" function involves creating a climate for ideas creation and filtration, and generating ideas.

So, how does all of this relate to creating knowledge? According to the Delphi Group, the top five reasons for implementing a knowledge management solution are: organizing corporate knowledge (63%), new ways to share tacit knowledge (39%), support for research and knowledge generation (31%), new ways to share explicit knowledge (29%), and "smart" tools to aid the decision maker (26%). The ability to create knowledge is related to The Delphi Group's third top reason for implementing a knowledge management solution, namely, to support knowledge generation.

Tacit to Explicit Knowledge

A key to creating knowledge is to tap the pool of tacit knowledge and convert it into explicit knowledge. Tacit knowledge refers to the "automatic" knowledge, that is the knowledge one uses without perhaps even realizing it. This may be the "subconscious knowledge." Explicit knowledge is brought out into the open and can be formalized into a knowledge repository, perhaps.

How does one elicit this tacit knowledge in order to represent it in a knowledge repository? This is the same situation that knowledge engineers faced in trying to acquire knowledge from a domain expert and then represent/encode it into an expert system's knowledge base. Various knowledge acquisition techniques have been used by knowledge engineers when constructing expert systems, and the same techniques also apply in trying to acquire this tacit knowledge and encode it in a knowledge repository.

Structured and unstructured interviewing techniques can be used to identify the tacit knowledge and externalize it. Limited information tasks or constrained processing are interviewing approaches in which limited information (or a constrained period of time) is given to the expert to see how the expert reasons and applies certain rules of thumb (i.e., heuristics — that could be part of the tacit knowledge) in solving a problem. The method of tough cases is also used to uncover some of these heuristics and tacit knowledge. Familiar tasks are scenarios that fall squarely within the domain of the expert, and the expert can comfortably apply these situations to his/her decision making process. Protocol analysis, a verbal walkthrough by the expert in stating aloud his/her decision-making process when solving a problem, has also been used to uncover some tacit and explicit knowledge. Other knowledge acquisition techniques include observation, simulations, and hypermedia-related approaches.

Some of this tacit knowledge may be uncovered through informal knowledge exchanges, such as meeting in the coffee area, exchanging conversations with colleagues, the "grapevine", and knowledge fairs or exchanges with employees. Organizations like Johnson and Johnson and the World Bank make great use of these "knowledge fairs" to encourage the sharing of knowledge and information between employees. Perhaps, through this dialogue, tacit knowledge will be revealed.

Converting Internalized to Explicit Knowledge

A related transformation that needs to be made is converting internalized knowledge into explicit knowledge. Internalized knowledge is related somewhat to tacit knowledge, but refers more to taking someone else's knowledge and then shaping it into one's own knowledge for better understanding by that person. Certain beliefs, values, truths, cultural factors, and the like may become part of the baggage associated with one's internalized knowledge. Some of this baggage may be important, but in other cases, some filter may need to be applied to weed out any biases affecting the purity of the knowledge item. The internalized knowledge will need to be made explicit so that others can share and apply this knowledge.

Converting Externalized to Explicit Knowledge

Externalized knowledge refers to the "outside" knowledge — that is, knowledge associated with the external environment such as through competitor intelligence, economic conditions, etc. Synthesizing knowledge from the external environment to create value internally to the organization is an important form of knowledge creation. A fairly new area within organizations is called CI for "Competitive Intelligence". Capitalizing on this competitive intelligence can strategically advance the organization.

Summary

These three conversion modes can help create knowledge and facilitate a knowledge sharing environment. Some knowledge creation techniques are shown below:

- Socialization
- Observation
- War stories (anecdotes of evidence)
- Shared story-telling
- Exchanging artifacts (documents, books, reports, files, software, memos, etc.)
- Simulation
- Metaphors, analogies
- Deduction, induction, dialectic reasoning, contradictions, paradoxes
- Brainstorming
- Prototyping and experimenting
- Face-to-face dialogue and group work
- Groupware, intelligent decision support, creativity/idea processing systems

If one follows a social theory of knowledge, the following could be said, as Bruce Gold points out:

- Knowledge management must be tied to an organizational purpose;
- Knowledge management must manage knowledge across a qualitative/quantitative continuum
- Knowledge management is about systems of knowledge as well as information and data
- Knowledge management is a socially/organizationally located activity
- Knowledge management must deliver knowledge at the level of specificity or abstraction appropriate for its users

According to Thomas Stewart's book, the new wealth of organizations is their intellectual capital. How can organizations increase their intellectual capital and become knowledge-creating enterprises? Investing in the organization's human, structural, or customer capital should hopefully increase the organizational intelligence of the entity. For example, surrounding oneself with smart and talented individuals should permeate an atmosphere of knowledge-seekers. Hopefully, a synergy is created which spreads through the rest of the organization. As this happens, new knowledge should ensue through the stimulation of novel ideas and through the combination of existing knowledge.

References

Glynn, M.A. (1996), Innovative Genius: A Framework for Relating Individual and Organizational Intelligences to Innovation, *Academy of Management Review*, Vol. 21, No. 4.

3 Capturing and Storing Knowledge

If one believes that tacit knowledge needs to be formalized and made explicit to be shared more easily, then an important part of the knowledge management process is to capture and store this knowledge into knowledge repositories. This creates an interesting question: How do you know when something is deemed "knowledge"? In fact, the author believes that "negative" experiences should also be part of the knowledge that's captured, as well as the "positive" lessons learned. Addressing the question, how is something validated to be declared "knowledge", vs. simply being data or information?

Knowledge is the capability to act, according to Hubert St. Onge of The Mutual Group in Canada. A panel of experts, as judged by their peers, could be assembled to help validate whether something is truly "knowledge" vs. information (patterned data) or data (dispersed elements). Alternatively, the organization may develop attributes of knowledge and rate the alleged pieces of knowledge against these criteria and attributes. Sample attributes of knowledge may be permanency, ingenuity, creativity, value-addedness, possible worth, and the like.

In the next section, we will discuss ways of capturing knowledge. Many of the techniques have evolved from the cognitive psychology and artificial intelligence fields.

Techniques for Capturing Knowledge

Various techniques have been applied over the years to capture knowledge. Some of these approaches will be highlighted.

Interviewing

The most typical method for capturing knowledge is through structured and unstructured interviews. Structured interviews follow a sequence of questions prepared in advance with some flexibility for deviation. Unstructured interviews are typically open-ended questions, not planned in advance. The skill of the interviewer is a key ingredient towards getting at the deeper knowledge vs. the shallow knowledge of the expert. Various interviewing methods can be used: the method of familiar tasks (typical scenarios faced by the expert), the method of tough cases (special cases faced by the expert), the constrained processing task (expert must make a decision within a constrained period of time), the limited information method (expert is given restricted information upon which to use for making a decision), and other interviewing techniques.

Protocol Analysis

Protocol analysis deals with verbal walkthroughs in which the expert reasons aloud through his/her decision making process without intervention of the interviewee. The expert describes the dead-end paths, as well as those that follow a positive direction, in reaching a decision.

Questionnaires and Surveys

Questionnaires, surveys, and web-based observation forms are alternative techniques for capturing knowledge. Lessons learned, best practices, and heuristic knowledge (rules of thumb) could be described on these forms to replace, or usually supplement, interviews.

Observation and Simulation

Some knowledge can perhaps be gained through observing an expert solving a problem in his/her work environment. In this technique, the knowledge engineer is a casual observer, watching the actions of the expert, firsthand, in the workplace. Computer simulation may also trigger the application of knowledge by the expert in running through various simulations. Role-playing could be used as a human simulation vs. computer simulation.

Learning by Doing

Perhaps through an apprenticeship, mentoring, or on-the-job training, the knowledge engineer may be able to pick up some insights and knowledge used by the expert, as the knowledge engineer learns directly from the expert (via teaching, doing, etc.). This master-apprentice relationship may uncover some of the knowledge that the expert applies. However, a line from an old movie comes to mind whereby the mentor tells his mentee — "Everything *YOU* know, I taught you. But I didn't teach you everything that *I* know!"

What to Capture?

Knowledge can take various shapes and forms. Knowledge may be expressed via the following sample mechanisms:

- Anecdotes
- War stories
- Case studies
- Lessons learned
- Best practices
- Failures and successes
- Heuristics (rules of thumb acquired through experience)
- Value-added relationships, regarding human capital, structural capital, and customer capital

By capturing these types of knowledge, organizations can begin to evolve into "knowledge organizations." Unfortunately, however, today's organization is typically not a knowledge organization. In fact, a survey of 200 Fortune 500 companies found that only 4 of these companies considered themselves a knowledge organization. A case study performed by the author illustrates this phenomenum.

Case Study

Ask almost any Chief Executive Officer what separates their company from their competitors. The typical response is "our knowledge." Many leading organizations are now discovering that they need to do a better job of capturing, distributing, sharing, preserving, securing, and valuing this precious knowledge in order to stay ahead, or at least survive, with their competition. Companies like Coca Cola, Sequent, Hewlett Packard, PriceWaterhouse Coopers, and others have established new positions within their companies to oversee and better manage knowledge in their organizations. This new position is often called the "Chief Knowledge Officer", and Knowledge Analysts (like at FedEx) assist the CKO in analyzing the knowledge processes within the firm in order to improve human performance.

The process of managing knowledge in organizations is referred to as "Knowledge Management" or "Enterprise Knowledge Management" (Liebowitz and Wilcox, 1997). Knowledge management is the process of creating value from an organization's intangible assets. The focus is to provide mechanisms for building the institutional memory or "knowledge base" of the firm to better apply, share, and manage knowledge across various components in the company.

Some companies are now being called "knowledge organizations" (Liebowitz and Beckman, 1998), as we have moved beyond managing "information" to managing "wisdom." Wisdom is the *application* of "knowledge" in the right settings. Without such knowledge or wisdom, companies might falter into social Darwinism where the weaker firms get eliminated. In order to best leverage knowledge or wisdom in the organization, companies are transforming into "intelligent organizations". With reengineering, downsizing, rightsizing, and other "ing's" in vogue, companies are looking toward maximizing their knowledge in organizations in creative and intelligent ways.

Organizations need to get on the information and knowledge superhighway in order to stay competitive (APQC, 1998). Direct marketing and improved distribution channels can be facilitated by the Internet to better target and serve customers. Additionally, many organizations are using intranets to build a knowledge repository (Liebowitz, 1997; 1998; 1999) within their organizations for improved sharing of knowledge and information.

The following is a case study examining the knowledge management activities of a leading US military organization specializing in strategic leadership. This organization has about 120 employees, 50 of which can be considered knowledge workers. It is argued that knowledge management is such an important topic that Information Systems (IS) curricula must include a course in this area.

Strategy, Approaches, and Processes

Based upon the survey results, the organization under study does not know how to measure the value of its intellectual capital. People, processes, and products/services are the most important knowledge carriers in its organization. The organization's overall strategic goals do not include knowledge management explicitly. A knowledge management initiative has been conducted for less than one year. Customer intimacy (focusing on providing "total" solutions for a well-selected group of customers) is currently the strategic emphasis within the organization. The most important knowledge management objectives in the context of the organization's business strategy are a combination of customer knowledge and internal know-how. The least important knowledge management objectives are acquisition of new knowledge from external sources, generation of new knowledge inside the organization, standardization of existing knowledge in the form of procedures/protocols, transforming individual (people's) knowledge into collective knowledge, and facilitation of the "reuse" and consolidation of knowledge about operations.

An "Intranet" (including groupware) is the major approach used in the organization for sharing and combination of knowledge. Lessons-learned analysis is the key approach used for the creation and refinement of knowledge. Lessons-learned inventories are used for the storing of knowledge in the organization. Each of these approaches is used at a business unit-specific level vs. an organizational-wide level.

About a year ago, the key knowledge-management-related activity in the organization was established to have a lessons-learned section in the organization's intranet. The focus was to share lessons learned with new employees and organizations. It is unclear as to what the bottom-line impacts of this practice area are at this point.

A second example of a knowledge-management-related activity used in the organization is an intranet repository of information for organization exercises. The senior leaders of the organization started this project about two years ago and it is an integral part of the organization's exercises. It has resulted in improved sharing of information and archiving of information and knowledge.

Culture

The decentralized operations of the organizational culture seem to support effective knowledge management. A lack of teamwork among subelements of the organization appear to be barriers to effective knowledge management in this organization. There has been little to no organizational buy-in and acceptance of knowledge management at the senior, middle, and supervisory management levels. There has been much buy-in and acceptance at the professional/knowledge worker staff levels. Currently, there are no specific training programs in place to support knowledge management. No incentives or reward system is in place to support knowledge management.

Technology

Information technology (IT) is used as an enabler in the organization to learn and innovate to do the job better, and to create new products and services. There is no "formal" WebMaster function. Intranet technology is used to support knowledge management efforts via shared documents/products and the gathering and publication of lessons learned/best practices. Internet functions are also used for knowledge exchange with suppliers and customers. Knowledge-based systems and centralized lessons learned/best practices databases are used in the organization as well for supporting knowledge management functions. More expert systems are being planned to support knowledge management, especially the use of intelligent agent technology over the intranet.

Business Outcomes

The organization currently does not capture, measure, and track the value of the organization's knowledge. It doesn't measure/track the (new) knowledge generation within the organization, nor is knowledge sharing measured within the organization. No measurement system is in place that shows how knowledge management affects the bottom line. No measurement is made of the value added and cost of knowledge management in the organization. The desired business benefits for the organization through the systematic management of knowledge and intellectual capital are improving (quality) products and services, innovation (new products and services), and improved (strategic/tactical) decision making.

Increased innovation in the organization is measured mainly by anecdotal evidence and evaluations/surveys. Business growth is measured by evalutions/surveys, and practice and process improvement is measured by anecdotal evidence in the organization. Increased customer satisfaction is measured by evaluations/surveys, and enhanced employee capability and organizational learning are measured by anecdotal evidence. Knowledge management isn't currently integrated with business processes or product processes in the organization.

Knowledge Survey of Valuating Human Capital in the Organization Under Study

A knowledge survey dealing with human capital was completed in order to determine how the organization contributes to each factor below.

Training and Education

- Formal training of employees: one week orientation to senior employees; usually one week every three years to other staff.
- Formal education of employees (i.e., degrees): senior employees have usually Master's degrees and many have Ph.D's.; most of the senior employees have been through senior service college; the other staff have mostly a high school education.
- Mentoring and on-the-job training: it does happen but is not formalized.

Skills

- Research skills: for senior employees, it's a definite plus.
- Entre- and intrapreneurship skills: minimal.
- Retention rates of employees: senior employees — high; other staff — fair.

Outside Pressures and Environmental Impacts

- Industry competition: for senior employees — not much competition; information systems/computer science-related employees — highly competitive; other staff — very little competition.
- Half-life of information in industry: senior employees — strategy, yearly; information systems staff — 6 months; other staff — 2 years.
- Demand and supply of those in the field: senior employees — demand low, supply medium; information systems staff — demand high, supply medium; other staff — demand low, supply high.

Internal and Organizational Culture

- R & D expenditures of the organization: very small in comparison to operations expenditures.
- Formalized knowledge transfer systems (e.g., lessons-learned databases or best practices guidelines institutionalized within the organization): lessons-learned section of the organization's intranet.
- Informal knowledge transfer systems (e.g., speaking with top management, secretaries and assistant to top management, attending organizational events, the "grapevine"): very informal, dependent on individual initiative.
- Interaction with customers and users: highly structured for selected individuals to perform most interactions with customers.
- Physical environment and ambiance (e.g., nice office, reasonable resources, etc.): great office, computers, and resources for performing jobs.
- Internal environment within the organization (e.g., reasonableness of demands by management placed on the employees, etc.): very reasonable.
- Short-term (2 to 4 years) and long-term (5 years or more) goals/prospects, from the employee's perspective, of the organization's viability and growth: organization's very short term goals (1 year) are known; no knowledge of short term or long term goals for the organization; assume current mission and goals will be the future goals.

Psychological Impacts

- Morale (attitude, benefits, compensation, conferences, travel, vacation time, etc.) of employees: senior employees — good; information systems staff — poor; other staff — average.
- Creativity and ingenuity of employees — good.
- Employee stimulation and motivation — good.

Ranking of Factors from 1 (Most Important) to 19 (Least Important) in Terms of Their Importance Towards Contributing to Human Capital Growth in the Organization

		For the top five factors, rate how well (Excellent, Good, Fair, Poor) the organization is satisfying them:
1.	Employee morale	Fair
2.	Employee stimulation and motivation	Good
3.	Short term and long term goals	Fair
4.	Formalized knowledge transfer systems	Poor
5.	Informal knowledge transfer systems	Fair
6.	Employee creativity and ingenuity	
7.	Formal education of employees	
8.	Formal training of employees	
9.	Mentoring and on-the-job training	
10.	Retention rates	
11.	Entre- and intrapreneurship skills	
12.	Internal environment within the organization	
13.	Physical environment and ambiance	
14.	Research skills	
15.	R & D expenditures of the organization	
16.	Interaction with customers and users	
17.	Demand and supply of those in the field	
18.	Industry competition	
19.	Half-life of information in industry	

Analyzing this Organization from a Knowledge Management Perspective

In analyzing this organization from the Delphi survey results, this organization is implementing some knowledge management activities (such as a lessons-learned section on their organization-wide intranet), but little has been done. Even with this lessons-learned repository, there is a "knowledge attic" approach being used whereby passive data collection and passive analysis and dissemination are used. The knowledge attic approach is the simplest of the techniques for developing a knowledge repository where the lessons learned, in this case, are entered by the individual workers in the organization and are accessed if people in the organization want to see some of these lessons learned. There isn't an internal "knowledge transfer department" within the organization that is checking if a "lesson learned" is truly a lesson learned (i.e., meeting certain criteria and guidelines), and this department isn't playing an active role in analyzing these lessons learned and sending them to appropriate individuals in the organization. Interestingly enough, a different arm of this global military organization (not the specific organization being studied here) has a "knowledge publishing" approach to building their lessons-learned knowledge repository. In this other part of the organization, there is a Center for Lessons Learned, which allows electronic submission, via the web, of lessons learned to this Center. Correspondingly, once these lessons are received, they are checked for appropriateness as a lesson learned, and then are sent by the Center to individuals in the organization who would find these lessons to be of value.

The organization being studied here doesn't seem to value its knowledge based upon senior level support. Very few formalized knowledge transfer systems have been developed within the organization. Strategic goals and objectives are not easily conveyed from senior management to those in the organization. Very little training and education is provided to employees in the organization to further develop their skills and expertise. Very little is being spent in research and development to expand the knowledge base in the organization. There is also little concern for valuating knowledge in the organization in order to measure its value added being provided to the organization. Knowledge is typically transferred via word of mouth, internal memos, and through the grapevine.

From an organizational climate and culture viewpoint, knowledge management would enhance the activities that are stressed by the organization — namely, various strategic exercises that are performed throughout the year in terms of wargaming and simulations. Sharing lessons learned and best practices from one year's exercise to the following year's exercise would greatly benefit and improve the exercises from one year to the next. This is being accomplished a bit, but more threaded discussions, on-line forums, and categorized lessons learned could greatly add to this experiential base. Additionally, a formalized knowledge transfer system could be developed in order to better convey the strategic, short-, and long-term goals of the organization from senior leadership to others in the organization, and to provide input from the various organizational members towards creating these goals and objectives.

The top five factors cited in terms of importance towards contributing to human capital growth in the organization were, in order: employee morale, employee stimulation and motivation, short- and long-term goals, formalized knowledge transfer systems, and informal knowledge transfer systems. It appears that only employee stimulation and motivation is rated as "good" in terms of how well the organization is satisfying their objectives. The other factors were rated as either "fair" or "poor" in terms of contributing to human capital growth in the organization. A slight paradox is occurring in the organization whereby employee stimulation and motivation is deemed higher than employee morale. One would think that the two factors would correlate well, but this may not be the case with this organization. This would have to be further investigated.

This organization certainly is not a "knowledge organization", due to its lack of knowledge management activities and processes throughout the firm. (As mentioned previously, a recent survey indicated that out of 200 Fortune 500 companies, only 4 of them thought of themselves as a "knowledge organization.") A knowledge organization is an entity that leverages and maximizes its use of knowledge in a value-added way, internally and externally, to the organization.

How to Store Knowledge?

Once knowledge is captured, it needs to be stored in a knowledge repository for distribution, retrieval, and access. Various approaches have been used to store knowledge.

One approach is the "knowledge attic" technique. This is the simplest form of a knowledge repository, in which knowledge is contributed by the employees in a passive way, and passive analysis and dissemination are applied whereby no entity analyzes the knowledge to disseminate to appropriate individuals in the organization who would benefit from its use. The knowledge attic approach is similar to having a corporate memory or repository sitting there for possible use.

Another approach for storing knowledge is the "knowledge sponge" method, which involves active collection but passive analysis and dissemination. Here, there is an active process to capture and store knowledge, but still there isn't an entity that analyzes and disseminates the knowledge to pertinent individuals in the organization.

The "knowledge publisher", a third approach, is passive collection but active analysis and dissemination.

The last major method to store knowledge is the "knowledge pump", which allows active collection, analysis, and dissemination.

References

American Productivity and Quality Center (1998), *Benchmarking Study on Knowledge Management for External Customers,* in cooperation with Inference Corporation, Houston, Texas.

Liebowitz, J. (Ed.) (1997), *Failure and Lessons Learned in Information Technology Management: An International Journal,* Cognizant Communications Corp., Elmsford, New York.

Liebowitz, J. (1998), *Information Technology Management: A Knowledge Repository,* CRC Press, Boca Raton, FL.

Liebowitz, J. (Ed.) (1999), *Knowledge Management Handbook,* CRC Press, Boca Raton, FL.

Liebowitz, J. and T. Beckman (1998), *Knowledge Organizations: What Every Manager Should Know,* St. Lucie/CRC Press, Boca Raton, FL.

Liebowitz, J. and L. Wilcox (Eds.) (1997), *Knowledge Management and Its Integrative Elements,* CRC Press, Boca Raton, FL.

4 Transforming Individual Learning into Organizational Learning

A critical element for successful knowledge management and for advancing the organizational intelligence of an entity is to be able to transform individualized learning into organizational learning. The notion of synergy, the whole is greater than the sum of its parts, is an essential ingredient for building organizational intelligence. The "collective" knowledge of an organization must be synthesized from the learning and key experiences of individuals throughout the company, as well as from the customers and stakeholders.

Part of this idea goes back to the "learning organization." A learning organization is an organization that has an enhanced capacity to learn, adapt, and change (Gephart, 1996). It's an organization in which learning processes are analyzed, monitored, developed, managed, and aligned with improvement and innovation goals. The essence of a learning organization, according to Martha Gephart (p. 38) is

- Continuous learning at the systems level: Individuals share their learning in ways that enable an organization to learn by transferring knowledge across it and by integrating learning into organizational routines and actions.
- Knowledge generation and sharing: Emphasis is placed on creating, capturing, and moving knowledge rapidly and easily so that the people that need it can access and use it quickly.

- Critical, systemic thinking: People are encouraged to think in new ways and use productive reasoning skills systemically in order to see links and feedback loops, and critically in order to identify assumptions.
- A culture of learning: Learning and creativity are rewarded, supported, and promoted through various performance systems from the top down.
- A spirit of flexibility and experimentation: People are free to take risks, experiment, innovate, explore new ideas, and generate new work processes and products.
- People-centered: A learning organization provides a caring community that nurtures, values, and supports the well-being, development, and learning of every individual.

So what are the different types of learning that can take place? According to Michele Darling, Executive Vice President of Human Resources at the Canadian Imperial Bank of Commerce (CIBC), there are four types of learning: *Individual learning* whereby the responsibility for continuous learning is placed on the individual; *team learning*, which involves task-focused employee groups that take responsibility for their learning; *organization learning*, which involves sharing team successes and achievements through learning networks; and *customer learning*, which involves the organization and everyone in it needing to know more about the customer and his/her evolving needs (Darling, 1996).

The process of organizational knowledge construction takes these four types of learning and incorporates them into a knowledge management and transfer system for access, use, manipulation, and distribution to employees, management, and customers. The way individual or local knowledge is "incorporated" into collective knowledge or organizational knowledge is central to knowledge management and organizational intelligence. Huysman et al. (1998) refer to organizational knowledge as practices, procedures, stories, technologies, collective opinions, paradigms, frames or references, etc., through which organizations are constructed and through which they operate. Huysman et al. stress that organizational knowledge is independent from the individual actor — that is, the organization learns only insofar as individual skills and insights become embodied in organizational routines, practices, and beliefs that outlast the presence of the originating individual. In this manner, knowledge management is a means through which organizational learning processes can be supported.

Huysman further points out that there are aspects of problematic learning that can handicap learning. These include:

- Filtered learning: Actors filter the knowledge that they exchange
- Egocentric learning: The organization interprets information from its own frame of reference
- Unbalanced learning: The organization learns only from a selected group of actors
- Autonomous learning: Actors do not learn from the organization

To reduce the risk of this handicapped learning, organizations can gain more information about the actors within the field and their interrelation with other actors. Additionally, internal learning within the organization and external learning between organizations must be conducted to enhance overall organizational intelligence.

The Knowledge Map at Monsanto Company

Monsanto Company is a leader in leveraging its collective intellect. One of the techniques the company uses as part of their knowledge management methodology is called the "knowledge map." The knowledge map involves two aspects: knowing how individuals learn and create knowledge, and knowing how teams made up of individuals collectively create insight. In terms of knowledge creation in individuals, there are four elements used (Junnarkar, 1997):

1. *Tacit-to-Tacit Knowledge Creation*: Face-to-face meetings, teleconferences, and video conferences are some of the common methods of exchanging tacit knowledge.
2. *Tacit-to-Explicit Knowledge Creation*: The most common form of capturing tacit knowledge and making it explicit is e-mail.
3. *Explicit-to-Explicit Knowledge Transfer*: E-mail, Internet, Intranet, Lotus Notes, distribution of printed documents, CD ROMs, etc. are examples of how IT has greatly enabled this type of knowledge transfer.
4. *Explicit-to-Tacit Knowledge Creation*: Successful sense making depends on one's expertise, experiences, mindset, and other attributes.

For understanding how teams interact, Monsanto looks at socialization (the dialogue or discourse that takes place among team members), externalization (the articulation of team goals), combination of insight (the collective intellect of the team), and internalization (the process of action). Knowledge maps take into consideration the processes and interactions at the individual as well as the team level (Junnarkar, 1997).

One way of transforming individual learning into organizational learning is to make the individual learning experiences readily available and accessible to those in the organization via compiling them into a knowledge repository. For example, case-based reasoning involves populating a case base of various situations-actions and then using similarity matching to retrieve the relevant cases for solving a particular type of problem. Help desk applications often use case-based reasoning for troubleshooting problems. Broderbund, one of the leading computer game manufacturers, uses case-based reasoning to allow their customers to solve problems with their computer games via the Web. It is an online version of their technical support whereby cases of problems and solutions have been populated in a case base, and then, through case-based reasoning, the customers can type in their problems and the relevant cases will be retrieved to tell the customer how to resolve their difficulties. This is the result of Broderbund having its technical support people indicate their individual learnings involving the problems and solutions relating to inquiries from the customers of computer games. This knowledge was formalized into a case base or a knowledge repository for access by the customers on line. This compilation of individualized experiences into an organizational knowledge repository added great value to their customer support and thereby increased customer loyalty to their products.

Another example of transforming individual experiences into organizational learning is through developing multiple cooperating expert systems whose contained knowledge can be shared by those in the organization. An expert system is a computer program that acts like a human expert in a well-defined task of knowledge. The heuristics (i.e., rules of thumb acquired by years of experience) are typically captured in the expert system's knowledge base. If these expert systems and associated knowledge bases are linked together via web-based, intranet, or knowledge sharing technologies, then they could become part of the organization's corporate memory and be accessible to those in and outside the organization.

However, making these individual experiences and lessons learned available to those in the organization doesn't guarantee by any measure that the organizational intelligence will grow and that the organization will learn collectively. Building a supportive culture to encourage organizational learning and growth is critical to developing a learning organization. This environment is stimulated through top management and senior leadership and emanates down through the organization. For example, CEO Bob Buckman of Buckman Laboratories used their knowledge transfer system, K'Netix, on a daily basis and would also send notes to selected employees and management in the organization if they hadn't used K'Netix in a while. Other companies have required, as part of the annual job evaluation review, that the employees access and apply knowledge in their organization's knowledge repositories in a value-added way. This helps to ensure that the organization as a whole is actively contributing their knowledge and applying it to, hopefully, create new knowledge and deposit it back into the knowledge repository.

To better understand how organizational learning can occur at the top levels of the organization, Clive Holtham and Nigel Courtney of the City University Business School in London performed a study on the executive learning ladder. Their stages of executive learning include:

1. At the lowest rung, the novice executive is primarily concerned with extracting information from data.
2. At the highest level, the executive is already an expert and develops further degrees of knowledge and insight through interaction with other experts.
3. The middle rungs place an emphasis on developing confidence and competencies through the sharing of experience — both good and bad — and by using case studies as a source of ideas.
4. At each layer, the dynamics of gathering data, acquiring information, developing knowledge and gaining insight take place.

From their study of executives at over 20 companies in the United Kingdom, Holtham and Courtney found that a variety of approaches to learning is required. The more experienced the executive, the fewer the number of stimuli required to trigger recognition and understanding. At the higher levels of the learning ladder, the contribution of expertise is as important as the acquisition of fresh information.

The learning process must also be dynamic. Dynamic learning by executives remains dependent on face-to-face meetings, according to Holtham and Courtney's study. This suggests the need for having informal communication, in addition to formal communication (e.g., reports, documents, knowledge repositories, etc.), as an important way to learn and increase organizational learning.

Metaphors speed up managerial learning. Senior executives used metaphors to bring an ad hoc group very quickly to a shared understanding of the issues and to make outcomes memorable. Junior managers require concrete examples to stimulate learning.

The study also found that knowledge-sharing behavior depends on the belief that sharing will lead to accrual. Other findings from this study were:

- Despite all the advances in IT-mediated communications, senior executives uphold face-to-face contact as the critical factor for successful development of managerial competencies in the application of IT for business benefit
- Although e-mail is prevalent in U.K. organizations, executives demonstrate their actions that asynchronous electronic communication across organizational boundaries has very low priority
- With experienced executives, the immediate impact of metaphors can and does speed up their unlearning of earlier theories and absorption of new

References

Darling, M., Building the Knowledge Organization, *Business Quarterly*, Winter 1996.

Gephart, M., V. Marsick, M. Van Buren, and M. Spiro, Learning Organizations Come Alive, *Training and Development Journal*, December 1996.

Huysman, M., M. Creemers, and F. Derksen, Learning from the Environment: Exploring the Relation Between Organizational Learning, Knowledge Management, and Information/Communication Technology, Proceedings of the Americas Information Systems Conference, AIS, August 1998.

Junnarkar, B., Leveraging Collective Intellect by Building Organizational Capabilities, Special Issue on Knowledge Management, *Expert Systems With Applications Journal*, Pergamon Press, Vol. 13, July 1997.

5 Combining, Transferring, and Distributing Knowledge

Once knowledge is captured and secured, it typically is combined with other knowledge, either housed in the knowledge repository or from outside users, to form new knowledge. Once this new knowledge is created, it is then stored in the knowledge repository and transferred to appropriate individuals in the organization. This chapter will take a look at ways for combining, transferring, and distributing knowledge.

According to David Vance and Jim Eynon at Southern Illinois University, a knowledge transfer function can be written as:

$$K1 = [(I \times W\text{system} \times BW\text{source}) + (K0 \times W\text{receiver})]$$

where:

$K1$ is the amount of knowledge held by the receiver
$K0$ is the amount of original knowledge held by the receiver
$W\text{system}$ is the warrant of the knowledge management system itself
B is a number $0 < B < 1$ that modifies the value of warrant assigned to the provider
$W\text{source}$ is the warrant of the knowledge provider
$W\text{receiver}$ is the warrant of the receiver
I is the target information to be transferred.

The goal of a knowledge management system is to maximize the increase in the knowledge of the receiver: $DK = K1 - K0$. Warrant is that quality "enough of which", along with truth and belief, produces knowledge.

According to Vance and Eynon, people who work in close proximity to each other (e.g., geographically located, the same or related discipline) are more likely to be able to share knowledge. For those reaching across geographical distance or in other fields, the knowledge management system must include a more explicit warrant in order to transfer knowledge. Vance and Eynon indicate that cultural, economic, and political realities for the knowledge sender may be vastly different from those of the receiver and thus, warrant must be more explicit in terms of its assumptions.

Even though informal communication may be used for transferring knowledge, it may be useful to convert these informal communications of knowledge into formalized knowledge via the knowledge repository. One reason for doing so is the game of "telephone" whereby the sender tells some information or knowledge to the receiver and then through word of mouth it is passed down to other receivers. This "telephone chaining process" may result in bias and distortion of the original message/information/knowledge because there is increased noise when multiple senders and receivers are used. Thereby, the original message or nugget of knowledge may be transformed into a slightly different message or knowledge and the knowledge may be transferred incorrectly or than otherwise intended. Again, the need for formalizing the knowledge into some artifact is an important part of the knowledge sharing and management process in order to increase the organizational intelligence.

Transferring knowledge usually occurs in various stages of the knowledge management process. However, before it can be transferred to an individual, it usually is assimilated and combined with other knowledge in order to create new knowledge. There are various ways that knowledge can be combined.

Combining Knowledge

How does one combine knowledge to create new knowledge? Here are some examples:

- The development of the cellular phone, through combining the knowledge of telephone and wireless communications

- Any invention like the zipper, the palm-top, genetic engineering, playing a song that you want on a tape at the push of a button, etc.
- The development of any new theories
- Other creative, ingenious developments of basic (and perhaps, applied) research

The idea of combining knowledge is to take existing knowledge and augment it in a new way or merge it with other knowledge to create new knowledge. This may also be called the "synthesis" step. Various tools and techniques could also be used to assist in this step. Idea processing systems, decision support systems, and groupware could be applied to enhance the synthesis step. For example, a decision support tool called "Expert Choice" could be used to help quantify subjective/verbal judgments for reaching a decision. Expert Choice follows the "Analytic Hierarchy Process" whereby pairwise comparisons are made to determine the importance of the decision criteria vs. the goal and then the alternatives vs. each criterion. Once these pairwise comparisons are entered by the user, the overall ranking of alternatives is calculated in their "synthesis" step. Idea processing tools to help in brainstorming and other techniques could also facilitate the combination of knowledge to create new knowledge.

As knowledge is accumulated, combined, and then generated, the organization's knowledge base should grow and the organizational intelligence should, hopefully, increase. For the organizational intelligence to flourish, knowledge needs to be distributed to appropriate individuals in the organization who could benefit and apply this knowledge. Distributing knowledge will be explained next.

Distributing Knowledge

Knowledge distribution involves sending knowledge internally and externally to those who could benefit from the use and application of the knowledge. Typically, there is an infrastructure within the organization whose responsibility is to disseminate the knowledge to pertinent individuals or groups. Instead of simply having a passive distribution mode where it is up to the individual workers to access the organization's knowledge repository, it may be preferable to have a knowledge management team in charge of analyzing the knowledge and distributing it to employees, management, customers, and stakeholders, as relevant. There are also techniques that could be used

Sharing Knowledge is Power Versus Having Isolated Islands of Knowledge Without Bridges

to assist in this process. Intelligent agents could be applied to analyze the knowledge, e-mail, Web pages, and the like and to disseminate appropriate summaries or individual pieces of information and knowledge to those who should best make use of it. Data mining and knowledge discovery techniques could also be employed to inductively look for trends, relationships, and possibly new knowledge and information from the organization's knowledge repositories. This is already being done effectively in the marketing and finance fields.

Online communities which share a common interest are also ways of sharing and distributing knowledge. These may mimic Centers of Expertise or Knowledge Centers or Communities of Practice in organizations. Members of these communities share their experiences, thoughts, information, questions/answers, and knowledge over the Web. For example, online communities exist for Parkinson's Disease, Down's Syndrome, and other health-related communities. Knowledge is distributed via the Web to members of these online communities.

Summary

Once knowledge is captured and secured, it needs to be combined, transferred, and distributed for new knowledge to be created and retrieved. By going through this iterative cycle, the organization should become smarter and should become more competitive in the marketplace.

6 Building a Continuous Learning Culture

In order for knowledge management strategies to be successful in an organization, a continuous learning culture needs to be built. If this supportive learning culture is developed, the organizational intelligence should also grow. In this chapter, we will take a look at cultivating a continuous learning culture in an organization.

Many people like Bob Buckman, CEO of Buckman Labs, have said that 90 percent of knowledge management is building a supportive culture. Developing a continuous learning work environment is crucial to the success of an organization. According to Rosow and Zager (1988), a continuous learning work environment is one in which knowledge and skill acquisition are essential responsibilities of every employee's job. Dubin (1990) later adds that it is supported by social interaction and work relationships, and relates to developed formal systems that reinforce achievement and provide opportunities for personal development. There is an emphasis on innovation and competition, both inside and outside the organizational context. Organizational members share perceptions and expectations that learning is an important part of everyday work life. Therefore, according to Schein (1985), continuous learning should be part of an organization's culture, as culture has been defined in terms of shared values and beliefs.

Tracey et al. (1995) feel that a continuous learning culture has several elements:

- Social support: the extent to which supervisors and coworkers encourage the acquisition and use of any new relevant skills and behaviors

- Continuous innovation: the extent to which an organization promotes ongoing efforts to be innovative and progressive
- Competitiveness: the extent to which an organization promotes an image of being the best in its field through high levels of individual performance

It has been found that a continuous learning culture can influence specific behaviors associated with a particular training program. According to Lundberg (1996), the nature of an organization's culture is extremely important for organizational learning. Without an appropriate culture, there is no repository for learning.

Knowledge Management is Mostly Culture and People, With Technology Thrown In

According to Holsapple and Joshi (1998), an organization's cultural knowledge resources impact participants' behaviors (e.g., knowledge sharing vs. knowledge hoarding). Leonard-Barton (1995) indicates that "values serve as a knowledge screening and control mechanism."

Understanding organizations as learning systems is an important element for organizational intelligence to increase. The typical characteristics underlying a learning organization are having an encouraging climate for members to improve; making professional development mandatory; extending stakeholder relationships to include customers, suppliers, and other important stakeholders; and continually reinventing itself through development and transformation (Garvin, 1993). Halal et al. (1998) believe that organizational learning cannot be instituted directly but only encouraged and nurtured through a supportive infrastructure and culture. According to Larry Prusak of IBM, when it comes to managing knowledge successfully, culture trumps all other factors (Halal et al., 1998).

Hubert St. Onge, Vice President for Strategic Development at The Mutual Group in Canada, believes that the corporate culture should be cultivated towards a knowledge environment. His key propositions are (St. Onge, 1998):

- The leadership and culture of a firm determine its capability to develop relationships with customers.
- An organization's ability to build customer relationships depends on the readiness of its leadership to engage in the development of the required organizational culture.
- Values measurement can inform us about the readiness of leadership to make the necessary shifts.
- Values reside at the core of an organization's tacit knowledge. Making these values explicit builds an effective platform for accelerating capability building.
- The place where leadership and culture come together is in the necessity of the enterprise to commit to a clear vision that aligns leadership, culture, and branding.

In a value-centered leadership survey performed by The Mutual Group, the customers prefer to (from a values point of view): partner with the organization to help them meet their aspirations; treat them with dignity and maintain a good relationship; provide them with knowledge and value-added information; demonstrate social responsibility through community involvement (St. Onge, 1998).

Organizational Memory

As a learning culture grows and develops, the organizational memory should also increase. According to Ackerman and Halverson (1998), organizational memory is both object and process. Memory is both an artifact that holds its state and an artifact that is embedded in many organizational and individual processes. This view encompasses more than most knowledge management's views of organizational memory, whereby knowledge management largely restricts itself to viewing organizational memory to repositories of experience "objects" that are reusable (Ackerman and Halverson, 1998). They use the distributed cognition approach in which they feel there is no such thing as a *single* organizational memory.

Many researchers have linked organizational memory with organizational learning. If an organization learns, then the result should be available later (Ackerman, 1998).

Case Study: A Knowledge Management, Organizational Memory Approach for Building a Continuous Learning Culture (Cantu-Ortiz and Liebowitz, 1998)

A major corporation involved in the construction of mines was very interested in applying knowledge management ideas to promote, support, and strengthen collaboration among the work groups that intervene in the installation process of the mineral processing plants called benefit plants. The benefit plants are used to recover the minerals of the rocks extracted from the mine. Once the benefit plant is installed, recovering of minerals consists of four main processes: grinding, milling, flotation, and filtering.

The decision-making process through the four stages is complicated due to several factors. First, although some of the equipment is of traditional use, there are several configuration alternatives at other levels as well as several connections among them. In addition, during the process, people with a great variety of backgrounds are involved: some are in charge of the plant design, others of the necessary equipment acquisition, others in charge of the relations and contacts with external suppliers, others responsible for the project, etc. As an example, several problems were mentioned that the plant is currently facing and that seem to be related to the installation process. Those problems mainly affect production by diminishing it and consequently having money lost. For example, excessive wearout in some equipment was pointed out, which as a side effect shortens its life span. Due to the difficulty in accessing the revision points, sometimes the production process had to be stopped in order to examine the equipment. Problems with the used/rebuilt equipment and problems in controlling some points of the process were also cited. It was also mentioned that the installation was not maintaining proper orientation because of a lack of integration of people in that area; some of the construction criteria were rigid; lack of information about the suppliers made it difficult to demand attention to equipment malfunctions; and excess of work by the person responsible for the project made his job difficult.

From the description above, some of the problems have to do with the difficulty in establishing appropriate communication among the participants in the process, as well as the use and recording of the good and bad experiences, in order to take advantage of them. The objective from this scenario was to have a computational system that provides communication and allows experiences to be registered, accessed, and applied easily, in parallel with the design of working methods that consider cultural aspects and assures the use of the knowledge management technology.

On one side, this project deals with organizational learning, and on the other, the distribution, sharing, and reuse of knowledge, including good and bad practices. A knowledge repository needed to be created in which experiences would be captured. This repository could be developed on the Web and be accessed using the company's intranet. In this way, people involved in the plant installation process could find out the convenience of any decision, register new cases, access lessons learned, etc. It could be given flexible search information services as well as the implementation of knowledge maps that allow people to check specific doubts about any area with the person known as the expert in that area.

One of the most critical points for ensuring the success of the system relies on the control and supervision of the recording of experiences. It is necessary that someone with enough authority determine when an experience or practice, in addition to being good or bad, is valuable. It is essential to have a group of experts validate the lessons learned and also have a knowledge management infrastructure within the organization for actively collecting, analyzing, and disseminating the knowledge in the knowledge management system to appropriate individuals affiliated with the company. There are several ways to approach this problem.

One is to use company personnel, where those regarded as experts in the different areas and processes that have to do with the installation of the benefit plans would be chosen to play the role of "judges." They would be approached for their best practices/lessons learned, and then they could refer others in those targeted areas. Data collection can be performed by having a Web-based observation/lessons-learned template/form made available to those affiliated with the company for entering their observations and lessons learned. A common template needs to be developed before data collection can commence. These Web-based forms could be completed by everyone in the targeted areas in the company, and they would be asked to enter their critical knowledge/best practices/lessons learned in their respective areas. Possible incentives may be provided by the company to encourage the initial development of these lessons learned. Then, a group of experts in the related areas would validate the lessons learned to check their quality and assess their worth.

Another way is to rely on external consultants who could be suggested by the company, and who are recognized as "experts" in the installation of benefit plans. They would periodically validate the experiences registered in the system.

Top management support and active involvement are needed to provide a motivate-and-reward system to encourage people to submit their heuristics and lessons learned and to use the knowledge in the knowledge repository in a value-added way. Building this supportive culture is critical to creating a knowledge sharing environment. A knowledge management infrastructure group, those maintaining the knowledge repository, should also play an active analysis and dissemination role in pushing the valuable nuggets of knowledge and lessons learned to those who could best benefit from these lessons from within the company. This approach will allow for active data collection and active analysis/dissemination.

There is another critical aspect that has to do with the use of the system. The use of the system could be a company policy for motivating and rewarding personnel who use the system correctly. The system could track when, how, and how much is used by each user and the same expert group that validates the experiences could evaluate the right use of the system based on the measures mentioned before. As an example, some companies like Andersen Consulting and Lotus require their employees to access and use the knowledge in their knowledge management systems as part of their annual job performance review. In the long run, if the system represents a real help for users, these users will be the true advocates of its use and this was the major challenge of the effort.

References

Ackerman, M. (1998), Augmenting Organizational Memory: A Field Study of Answer Garden, *ACM Transactions on Information Systems*, Vol. 16, No. 3, Association for Computing Machinery, July.

Ackerman, M. And C. Halverson (1998), Considering an Organization's Memory, Proceedings of the CSCW '98, Seattle, WA, Association for Computing Machinery.

Cantu-Ortiz, F. and J. Liebowitz (1998), Knowledge Management for the Installation and Operation of Benefit Plans, Working document, ITESM/UMBC, November.

Garvin, D., Building a Learning Organization, *Harvard Business Review*, Vol. 71, No. 4, July–August 1993.

Halal, W., M. Kull, J. Liebowitz, and J. Artz (1998), Report on a Study of Organizational Intelligence, Funded by the National Science Foundation, George Washington University.

Holsapple, C. and K. Joshi (1998), In Search of a Descriptive Framework for Knowledge Management: Preliminary Delphi Results, Kentucky Initiative for Knowledge Management, University of Kentucky, Lexington, KIKM Research Paper No. 118, March.

Leonard-Barton, D. (1995), *Wellsprings of Knowledge*, Harvard Business School Press.

Schein, E. (1985), *Organizational Culture and Leadership*, Jossey-Bass.
St. Onge, H. (1998), Cultivating Corporate Culture Towards a Knowledge Environment, Proceedings of the New York Business Information Conference, TFPL, Inc., November.

7 Culture: The Key Ingredient

According to Michael McGill's and John Slocum's book, *The Smarter Organization*, corporate culture was to companies in the 1980s what the "one best way" had been in the prior 100 years. By focusing on culture, a fundamental understanding of what the company valued could be achieved. In the 1980s, the culture-change programs focused on determining the shared values, then using management practices to increase understanding across the company. McGill and Slocum point out that any strong culture company (like GM) will be limited in its capacity to learn by the very nature of its culture. Employees are to use their understanding of the company culture to relate to their jobs and guide their behaviors. McGill and Slocum add that a company involved in promoting its own culture is unlikely to be open to enhancing and expanding experiences in the way that learning requires; it cannot get outside the understanding box!

The key to success for many organizations like Home Depot, 3M, Southwest Airlines, the San Diego Zoo, and the like is the commitment to learning from every aspect of the organization's experience. As we learned in the last chapter, creating and fostering a climate that promotes learning will ultimately give a sustainable strategic advantage to the organization.

According to Hayduk (1998), several factors are attributed to the development of a culture within an organization. These include:

- Unique national cultures of the employees
- Business environment of the company (the reality in which the business operates, which is dependent on its products, competitors, cus-

tomers, technological dependence, and the level of government influence)
- Types of employees hired into the organization
- Dynamic relationship between the business environment and the employees hired into the organization
- Introduction of new processes into an organization
- Interaction between the employees and the systems, structures, and processes established by management

Hayduk (1998) further emphasizes that changing culture for the implementation of a knowledge management system, structure, or process within the organization is dependent upon a number of elements. First, changes in systems, structures, and processes needed to develop a knowledge sharing environment must be embraced by management. Support of a new system is through both communications and behavior of management. Management should also seek the behavior of employees that will complement changes in systems, structures, and processes that promote the sharing, adaptation, and application of corporate knowledge. Incentive systems to support the behavior they are seeking need to be developed. Additionally, the acceptance of opinion leaders (and users) within the organization will have significant bearing on the success of a knowledge management system. Providing positive alternatives to the old systems and designing an employee base that promotes the new culture should help in creating a changed culture.

The American Productivity and Quality Center has performed a number of benchmarking studies on knowledge management. One of the studies indicates that culture and behaviors are the key drivers and inhibitors of internal sharing (O'Dell and Grayson, 1997). The real issues are: how can people be motivated and rewarded for sharing and what can leaders do to help establish and reinforce a supportive culture? Another knowledge management study by APQC (1996) found the following common factors whose cultures have become more supportive of knowledge management (and thus growing the organizational intelligence of the entity):

- Leadership, leadership, leadership
- Make a big deal of successes (organization rather than individual), even if you have to go back and reframe an earlier organizational success as an example of successful knowledge management and transfer

- Provide the tools necessary for people to find the information and knowledge they need themselves
- Clarify the business case and value proposition in concrete terms

At Arthur Andersen, for example, there were two simultaneous approaches to creating a culture more supportive of knowledge management: the bottom-up approach creating a sense of community and trust among people, and a top-down approach creating the norms, standards, and overall value statements that are used by top management to drive strategies and (tolerated and not tolerated) behavior (APQC, 1996).

The ability of the organization to learn is an essential aspect of organizational intelligence. To do this, employees and management must be able to "learn how to learn" and be adaptable to changes in the marketplace. The "learning organization" notion should be transmitted through the organizational culture. The organizational culture is then made up of shared values, norms of behaviors, symbols and symbolic action. According to Beckman (1998), dimensions of values and beliefs are

- Truth, openness, reality
- Courage
- Self-management
- Conflict resolution
- Cooperation
- Diversity
- Decision making
- Participation
- Compensation
- Motivation
- Sponsorship
- Mistake-tolerance
- Authority
- Creativity
- Rituals, symbols, myths
- Consistency
- Fairness
- Risk-taking
- Teams
- Empowerment

- Collaboration
- Power
- Sharing
- Partnering
- Information access
- Commitment
- Risk-tolerance
- Competence
- Innovation
- Acceptable behaviors
- Change
- Focus, attention

According to Karl Wiig (1996), President of the Knowledge Research Institute, positive cultural factors include:

Positive Cultural Factors	Examples
Understanding factors	Clear mission statement — aggressive company goals which are conducive for knowledge management
Management practices — culturally internalized	Anybody can talk to anybody else — proactive management reacting to changing environment and industry
Operational practices — culturally internalized	Strong teaming culture which supports exchange of ideas
	Cross-functional execution of business initiatives
Role models and leadership	Effective champions who promote change, help teams withstand outside distractions
	Very strong leadership support from Chairman and CEO
Cultural driving forces	Openness and honesty, sincere service attitude towards membership
	High trust culture for shared learning

Cultural impediment examples include (Wiig, 1996):

Cultural Impediment Factors	Examples
Cultural driving forces	"It's not my job"
	Knowledge hoarding for personal/professional gain
Values	Concentration on financial goals
	"Not invented here" syndrome; "not invented here" thinking; too much "not invented here"
Beliefs	Management perception that organization possesses knowledge and people are expendable
	Innovation is highly valued and this detracts from using other's knowledge because their innovations will not be seen as "original"
Understanding-related factors	Lack of broad understanding of enterprise direction and strategies
	Knowledge management is perceived to be more work

To be truly effective, knowledge management must be accomplished by organizational change (Marshall et al., 1996). Marshall et al. (1996) in their study of risk management groups indicate that the firm needs to understand what knowledge it has and seek out the knowledge it needs; organizational knowledge should be transferred to those who need it in their daily work; organizational knowledge should be accessible to those who may need it as events warrant; new knowledge should be rapidly generated and made accessible throughout the organization; controls are developed to embed the most reliable and robust knowledge; organizational knowledge is tested and validated periodically; and the firm should facilitate knowledge management through its culture and incentives. Examples of incentives include awards (nonmonetary) and recognitions, bonuses and other monetary rewards, personnel evaluations and promotions, special focus meetings, and general communication approaches (Wiig, 1996).

Chrysler developed an Engineering Book of Knowledge (EBOK) concept involving continuously capturing and collaborating on the experience gained through the product development cycle (Ernst and Young, 1997). Through this EBOK, a knowledge sharing concept emerged at Chrysler whereby the EBOK contributed to learning (formal and informal) at Chrysler. This in turn developed into knowledge and then into insight. From the insight, innovations were made which accelerated performances, these new lessons learned were then documented within EBOK, and the cycle continues.

References

American Productivity and Quality Center (APQC)(1996), Knowledge Management Study: Final Report, Houston, Texas.

Beckman, T. (1998), Knowledge Management Seminar Notes, George Washington University, June.

Ernst and Young (1997), *Chrysler Corporation's Engineering Book of Knowledge*, Center for Business Innovation, Cleveland, OH.

Hayduk, H. (1998), Organizational Culture Barriers to Knowledge Management, America's Information Systems (AIS) Conference Proceedings, Baltimore, MD, August.

Marshall, C., L. Prusak, and D. Shpilberg (1996), Financial Risk and the Need for Superior Knowledge Management, *California Management Review*, Vol. 38, No. 3, Spring.

McGill, M. and Slocum, J. (1994), The Smarter Organization: How to Build a Business that Learns and Adapts to Marketplace Needs, John Wiley & Sons, New York.

O'Dell, C. and C.J. Grayson (1997), *If We Only Knew What We Know: Identification and Transfer of Internal Best Practices*, American Productivity and Quality Center, Houston, Texas.

Wiig, K. (1996), Ensuring That We Capitalize on the Use of Knowledge, Knowledge Research Institute, Arlington, Texas.

8 Developing a Knowledge Management Capability

An important part for building the organizational intelligence of an entity is to bridge individual learning and organizational learning. As the previous chapters have explained, one way of doing so is through developing a knowledge management capability within the organization.

In order for an organization to learn, there are several important requirements, according to the British Petroleum experience in organizational learning (Alworth et al., 1998):

- Define specific goals linked to the bottom line
- Access a reservoir of experience to profit from past success or failure
- Ensure actions are directed at meeting goals
- Accurately monitor performance
- Remember how decisions were made
- Improve an accessible reservoir of experience

To implement these requirements, various organizations have used a variety of knowledge management strategies. According to O'Dell and Wiig (1997), there are six major knowledge management strategies. The first is to use knowledge management as a business strategy. This would be the most comprehensive and enterprise-wide strategy, where knowledge is a product to be sold. A second strategy is to have a transfer of knowledge and best practices. In this strategy, systemic approaches to knowledge reuse and transfer are used whereby knowledge is obtained, organized, repackaged, and distributed. There is a need to reward suppliers and users. A third strategy is customer-focused knowledge, whereby customer knowledge is captured, developed, and transferred. This involves understanding the customer's needs, preferences, and business, and bringing knowledge of the organization to bear on customer problems. A fourth strategy is personal responsibility for knowledge. In this approach, one identifies, maintains, and expands one's own knowledge and also understands, renews, and shares knowledge assets with others. A fifth strategy is intellectual asset management. This involves managing structural knowledge assets and best practices, and renewing, organizing, evaluating, and protecting market assets. A sixth strategy is innovation and knowledge creation. This involves basic and applied research and development, and experimentation.

According to Beckman (1998), management consulting firms are among the leaders in knowledge management implementation. Their business is "selling knowledge" in such forms as results (best practices, information technology, plans, assessments), methodologies (reengineering, process improvement, systems life cycle, knowledge management), consulting services, and training. For knowledge management, most management consulting firms focus on capturing lessons learned from previous engagements, learning to improve bidding proposal techniques from past experiences and competition, and storing, distributing, and accessing methodologies and training materials.

Davenport and Prusak (1998) talk about leveraging existing management approaches. In terms of leading with technology, a technology infrastructure is necessary, but insufficient in knowledge projects. For success, focus on organizational/cultural issues. For leading with quality/reengineering practices, redesign business system components and this approach may be worthwhile if previous reengineering efforts were successful. For leading with organizational learning, concepts involve cultural and behavioral issues. Often, there is too little emphasis on knowledge and technology, and often the strategy is too conceptual and abstract for major impact. For leading with decision making, apply lessons learned to improve decision-making and value knowledge by calculating its costs when absent. For leading with accounting, create an accounting system for knowledge and intellectual capital. For leading with corporate culture, choose a management style consistent with the firm's culture and adopt multiple approaches for knowledge management to give this discipline broad-based institutional support.

At Andersen Consulting, there are well over 200 knowledge management jobs. Knowledge management functions are currently divided among several roles across knowledge managers. Andersen Consulting has "knowledge integrators" who are sufficiently expert in a particular domain to determine what knowledge is most valuable and synthesize it. There are also "knowledge administrators" who work at capturing, storing, and maintaining knowledge that others produce (Davenport and Prusak, 1998; Beckman, 1998).

At American Management Systems (AMS), a large consulting firm, knowledge centers were established in the company's core competencies. These knowledge centers were the basis for knowledge-based communities of practice. There are about 100 to 150 people in each knowledge center. They must be nominated by the business unit manager, and must commit to making a contribution to the knowledge repositories by a research paper, lessons learned, or a deliverable with value added description. This contribution must be reusable, and the knowledge center affiliates must provide expert feedback within 12 hours of request. The knowledge centers sponsor special interest groups around topics like Y2K, Lotus Notes, etc. At AMS, these knowledge centers have transformed AMS from a verbal to a written sharing culture. Items reused are tracked (the top ten for each quarter, with the author's name, are highlighted). There is an award for the most reused item in a year, and conferences/contests to allow face-to-face exchanges. Successes are celebrated through best practice awards for creation and reuse of knowledge assets. Serious anecdote management is also conducted to keep track of stories with significant dollar punch lines.

At The World Bank in Washington, D.C. (Denning, 1998), their Knowledge Management System consists of all the basic things (development statistics, directory of expertise, help desks, knowledge management systems on-line, etc.). However, a heavy emphasis is placed on networks of people across geographical areas and areas of competence (poverty reduction and management, environment, rural development, etc.) who facilitate the sharing of knowledge. These "communities of practice" lie at the heart of the system. A community of practice for The World Bank is a group of professionals, a common class of problems, a common pursuit of solutions, themselves embodying a store of knowledge. Within the broader communities of practice, there are "thematic groups", which are smaller and help knowledge sharing among the whole community. Thematic groups are formed and re-formed according to the need. Around 40 thematic groups currently exist and may include task teams. The main activities of thematic groups are development and maintenance of knowledge content and quality assurance; outreach, promotion, partnering, and dissemination; and support for task teams. The World Bank strongly believes in needing people-to-people contact, thus they have arranged for Knowledge Fairs to get people together to talk about their common areas.

In an ongoing research project report (Abell and Oxbrow, 1999), "Investigation of Underpinning Skills for Knowledge Management: Training Implications," funded by the U.K. Library and Information Commission, several knowledge management trends seem to be emerging. First, most organizations talk about the need to identify achievable knowledge management activities that will have an impact on the business and to concentrate on those. Second, knowledge management is no longer confined to very large commercial corporations. Both public and private organizations are now considering ways of benefiting from some of the knowledge management concepts. Third, there is no blueprint for knowledge management. Knowledge management is a home grown activity which has to reflect the organization, its unique place in the market, and its culture. Fourth, the importance of commitment from the very top of the organization is essential for knowledge management to have a fighting chance. Fifth, the role of the CKO (Chief Knowledge Officer) is becoming more clearly defined as a leadership role, generally time limited, to provide the necessary focus and drive to steer a change management through the crucial stages. Sixth, knowledge management labeled roles are more likely to emerge as a realignment of an existing role, become an additional role for an existing post, or if identified as a new role, to be filled by existing staff. However, there are a number of new roles being coined such as knowledge analysts, knowledge architects, knowledge managers, knowledge brokers, knowledge auditors, knowledge navigators, knowledge guardians, and the like.

According to The Delphi Group 1998 survey on knowledge management implementation, in which 575 worldwide responses were received from over 18 industrial sectors, the status of knowledge management roles within organizations was the following:

Role	Status of Role within Organizations
Chief Knowledge Officer	12% already have it in place; 12% will within a year; 18% will within the next 2–3 years; 59% are unlikely ever to have such a position.
Knowledge Manager	15% already have it in place; 18% will within a year; 22% will within the next 2–3 years; 46% are unlikely ever to have such a position.
Knowledge Analyst	21% already have it in place; 19% will within a year; 21% will within the next 2–3 years; 40% unlikely.
Knowledge Architect	19% already have it in place; 17% will within a year; 16% will within the next 2–3 years; 47% unlikely.

In a study on knowledge management by the American Productivity and Quality Center (1996), the responsibility for knowledge management activities resided:

	Primarily in a Centralized Function	With the Business Units	In the Business Units with Support from a Centralized Function
Arthur Andersen			X
Chevron		X	
Dow Chemical			X
Hughes S&C			X
Kaiser Permanente		X	
National Security Agency		X	
Price Waterhouse			X
Sequent			X
Skandia			X
Texas Instruments			X
USAA		X	

New Positions in the Knowledge Management Field

Chief Knowledge Officer

Director of Intellectual Capital

Knowledge Management Director

Director of Shared Lessons Learned

Chief Learning Officer

Starting a Knowledge Management Program at Umax (Wiig, 1995)

In order to manage explicit knowledge-related work within Central Services (CS) at Umax, several areas needed to be coordinated:
- Cumulating work products in "knowledge repositories"
- Coordinating explicit knowledge activities such as knowledge analysis and knowledge transfers to points-of-action to ascertain that they fit together into a cohesive mosaic that supports Umax's overall objectives
- Ascertaining that sufficient resources are available to competently perform needed work
- Coordinating knowledge-related work with other entities (training, etc.)
- Selecting appropriate knowledge-related tools and approaches that have the broadest applicability without sacrificing specific needs
- Maintaining a perpetual overview of scopes, progress, and deliverables from knowledge-related work
- Building needed knowledge through acquisition, learning, and other means

References

Abell, A. and N. Oxbrow (1999), Investigation of Underpinning Skill for Knowledge Management: Training Implications, Interim Report, TFPL, Inc., London, January.

Alworth, C.D., E. Frost, and F. Kessler (1998), A Practical Approach Bridging Individual Learning and Organizational Learning: A Look at Organizational Learning in the E&P Industry, Proceedings of the 2nd International Conference on Practical Aspects of Knowledge Management, Basel, Switzerland, Oct. 29–30, U. Reimer (Ed.).

American Productivity and Quality Center (1996), Knowledge Management Study: Final Report, Houston, TX.

Beckman, T. (1998), Knowledge Management Seminar Notes, ITESM, Monterrey, Mexico, June 24–26.

Davenport, T. and L. Prusak (1998), *Working Knowledge*, Harvard Business School Press, Cambridge, MA.

Denning, S. (1998), The World Bank: Knowledge Management, Presentation from the 1998 Conference on Knowledge Management and Organizational Learning, The Conference Board, New York, April.

O'Dell, C. And K. Wiig (1997), *Knowledge Management Study*, American Productivity and Quality Center, Houston, TX.

Wiig, K. (1995), *Knowledge Management Methods*, Vol. 3, Schema Press, Arlington, TX.

9 Assessing Knowledge Management Through a Knowledge Audit

Many organizations may not have a good feeling for how well they are doing knowledge management, determining the gaps, and developing improved approaches for knowledge management. To facilitate this process, a knowledge audit can be conducted. According to ONLINX Research Inc., a knowledge audit is a review of the firm's knowledge assets and associated knowledge management systems (including the human capital, structural capital, and customer capital). It's a systematic and dispassionate review of the adequacy and integrity of important organizational assets and systems. At ONLINX, their knowledge audit approach is to first assess the strategic position of the firm (identify and clarify the true sources of competitive advantages for the firm — namely, core competency areas and associated strategic advantages); construct and assess the value network for each important business and associated critical success factors (also citing critical knowledge functions); map associated competencies and skills and assess by interviews, surveys, and other forms of research and data gathering (e.g., competitive benchmarks).

According to Peter Smith of ONLINX Research Inc., the value of a knowledge audit is to:

- Demonstrate exactly where value is being created through human and structural capital
- Highlight where leverage can best be applied through improved knowledge sharing and organizational learning

- Help prioritize projects for improving knowledge management practice
- Demonstrate firm capabilities to shareholders and other stakeholders
- Act as a key component to strategic planning for any knowledge-based enterprise
- Act as a key adjunct to due diligence and business planning in mergers, acquisitions, strategic alliances, venture capital, and new company formation

The knowledge audit team needs to determine the objectives of the audit. Are you looking for knowledge-based opportunities in the market? Is your focus knowledge flows, stores, and sinks in the organization? Is the audit to check compliance with standard operating procedures? Are you searching for ways to leverage internal processes using knowledge? Will you focus on knowledge objects or are you interested in cultural barriers? Is the audit strictly for gathering baseline data (descriptive) or is there a prescriptive element? Is the audit part of a larger business process reengineering project? Does the mandate include all stakeholders, suppliers, customers, stockholders, etc.? Where does the learning take place that generates the knowledge? How fast is the learning? What are the key leverage points in the learning process?

To help identify these goals and objectives, The Delphi Group's survey identified the top reasons why companies are implementing knowledge management solutions. They are: organizing corporate knowledge (63%); new ways to share tacit knowledge (39%); support for research and knowledge generation (31%); new ways to share explicit knowledge (29%); "smart" tools to aid the decision maker (26%).

One reason for doing a knowledge audit and performing knowledge management is to capitalize on knowledge reuse. Knowledge reuse provides for the capture and reapplication of knowledge artifacts (episodes in memory, stories, relationships, experiences, rules of thumb, and other forms of knowledge acquired by individuals or groups). Knowledge reuse relies as much on the use of negative experiences, flawed reasoning, or wrong answers as on correct results.

According to TFPL, Inc., a consulting firm based in England, their knowledge audit approach is:

- Identify the information needs of the organization, the business units, and individuals

- Identify the information created and assess its value
- Identify the expertise and knowledge assets
- Identify the information gaps
- Review the current use of external and internal information sources
- Map information flows and bottlenecks within those flows
- Develop a knowledge map of the organization indicating appropriate connections and collections

These tasks will enable the design of content maps for intranets, knowledge management strategies, and information strategies. Monsanto Company developed a knowledge management architecture which includes a learning map that identifies questions answered and resulting decisions made, an information map that specifies the kind of information that users need, and a knowledge map that explains what users do with specific information (conversion of information to insight or knowledge).

Essentially, according to TFPL, Inc., you are trying to identify how the organization can make better use of its intellectual assets and best leverage and share its knowledge internally and externally. With the audit, there should be an independent team; a mix of interviews, questionnaires, discussion groups, and focus groups; a number of people in "central" positions interviewed; a detailed questionnaire sent to all staff; and post-audits performed periodically.

Many organizations should perform a knowledge audit first. Factors to be considered are: What information do people need to do their jobs? What is the function of the information? Who holds that knowledge now? Who needs it? When? How can that information be made substantially more effective? What connection should you make between documents and work flow? Do you really want document management tied to work flow, processes, and business process reengineering? What precisely is the nature of your knowledge resources?

Teltech's knowledge audit process begins at the top, identifying the client's key decision-making areas and tasks, and drills down to evaluate the types, level, and location of information and knowledge required to support those decisions. Then a gap analysis identifies information/knowledge weak areas. Teltech's knowledge audit takes 3 to 4 weeks at one location with 200 to 600 employees. They assess knowledge-related behavior (e.g., how receptive is a client's culture to accessing information electronically?).

An important element of the knowledge audit is to be able to identify and share critical knowledge. Such questions as what is critical knowledge, where can we find this knowledge, how does the enterprise architecture enable the business to leverage knowledge across the corporation, and related questions should be asked as part of the knowledge audit. The knowledge auditors will need to focus efforts in promoting and enabling organizational and architectural changes while understanding the roles of a wide range of contributing technologies.

Appendix A describes an intelligent agent architecture for a knowledge management system for multimedia designers. Appendix B shows a knowledge audit instrument developed by the University of Manchester in England. Appendix C is a survey developed by the American Productivity and Quality Center, the European Foundation for Quality Management, and the Knowledge Management Network. Appendix D is a survey to look at an organization's human capital assets.

10 Augmenting Organizational Intelligence

Throughout the book, we have discussed how organizations can increase their "organizational intelligence" via knowledge management endeavors. As Governor Glendening of Maryland, in his state-of-the-state address on January 21, 1999, indicated:

> "...The future of our world is fast becoming knowledge based...Nowhere is the emphasis on knowledge more pronounced than in the job market...In the 21st century, the greatest skill will be the capacity to acquire new skills. In the new economy, acquiring knowledge will be a lifelong journey, not just a destination to be reached."

Echoing this statement and those by others, we have entered the knowledge age and knowledge industry. In order for companies to sustain their competitive edge, it is increasingly important that they replenish their intellectual capital and knowledge assets in their organization in order to be competitive in the marketplace. Most organizations currently are NOT "knowledge organizations". They have not totally embraced the concepts of knowledge management and have not maximized their use of knowledge internally and externally to their customers, stakeholders, and shareholders.

So, how can organizations increase their intelligence and rise to the level of a "knowledge organization"? Here are some strategies that may be useful:

- Invest in education and training of the firm's human capital

- Develop knowledge repositories for preserving, sharing, and distributing knowledge
- Provide a "motivate and reward" system to encourage employees and management to contribute to the organization's knowledge repositories and use this knowledge
- Consider evaluating annually each member of the firm on the quality and quantity of knowledge contributed to the firm's knowledge bases as well as the organizational knowledge used and value-added results by that firm member
- Develop methodologies for managing and structuring the knowledge in the knowledge repositories
- Consider having knowledge fairs/exchanges to encourage informal communication sharing among communities of practice
- Develop Centers of Expertise or Corporate Knowledge Centers with associated affiliates in the company in core competency areas
- Provide an infrastructure of individuals whose main job is to manage the creation, development, and maintenance of the knowledge management systems
- Adapt to the changing competitive environment by forming project teams based on the employee knowledge profiles
- Integrate knowledge management within the strategic goals of the organization

In the Halal et al. (1998) study on organizational intelligence, five key factors were identified that help comprise organizational intelligence. These were organizational structure, organizational culture, stakeholder relations, knowledge management, and strategic processes. In this report, those organizations exhibiting "low intelligence" would typically have a centralized structure, bureaucratic culture, conflicting stakeholder relations, isolated knowledge management, and authoritative strategic processes. Those organizations with "high intelligence" would have a decentralized structure, entrepreneurial culture, cooperative stakeholder relations, integrated knowledge management, and participative strategic processes.

In the same study (Halal et. al, 1998), the following functions were identified as related to organizational intelligence:

Function	Organizational intelligence
Measurement	Organizational intelligence quotient (if it could be determined)
Information technology	Organizational IT systems

Function	Organizational intelligence
Structure	Network of business units
Subjective filter	Organizational culture
External linkages	Stakeholder relations
Knowledge store	Knowledge management
Strategy formation	Strategic processes
Direction	Leader
Guidance	Mission
Decision-making	Strategy
Covert system	Information organization
Routine decisions	Policies and procedures
Knowledge gain (single loop learning)	Training and action
System improvement (double loop learning)	Organizational change

About 50% of the fastest growing companies in the U.S. are knowledge-intensive organizations, via selling the knowledge and the know-how of their employees rather than manufactured products or providing services (Mentzas and Apostolou, 1998). Management consulting firms are typical examples of knowledge-intensive organizations, and one would think they would be among the leaders of "knowledge organizations." However, according to a survey (Reimus, 1996) on knowledge management activities of consulting firms, about 60% of the consultancy firms maintained no active "best practices" database; less than 25% utilized the Internet to support a basic range of internal activities; and the actual mechanisms and processes in place for managing acquisition, screening, and selection of best practices at many consultancies were largely informal. Mentzas and Apostolou (1998) did a comparative analysis of experiences in managing corporate knowledge in consulting firms and concluded the following:

- The success of the knowledge management efforts was largely due to the creation of a knowledge-friendly culture in the organization
- A crucial element for a company to establish a positive knowledge culture is the type of people that a firm attracts and hires
- Knowledge management projects benefit from senior management support
- Clarity of purpose and vision is a critical factor with knowledge management efforts
- The right blend of information technology and organizational infrastructures is very important for knowledge management efforts to succeed

So why should companies be interested in knowledge management for increasing their organizational intelligence? According to Mahe and Rieu (1998), enterprises sometimes have to preserve knowledge, have suffered knowledge losses, need a large amount of knowledge, or need to lose knowledge. Knowledge preservation is needed in high technology sectors to avoid the serious danger that a loss of mastery of these technologies would represent. The problem of loss of knowledge is typically recognized after the fact, where vital expertise may have left the company through retirements, layoffs, transfers, departures, and the like. The greater need for knowledge is typically used to master the increased variety of products. The need to lose knowledge may be necessary to increase one's innovative capacity — that is, to sometimes forget what we know and start from scratch to design a new product.

In order for organizations to better leverage learning, Davenport and Prusak (1998), advocate the following:

- Think about organizations as a "system"
- Build and facilitate communities of learning and practice
- Focus on issues of personal development and "mastery"
- Create less hierarchical, more "self-organizing" organizational structures
- Plan with the use of scenarios

From the 1996 Knowledge Management Study, conducted by the American Productivity and Quality Center, several key findings emerged. At the time of the study, the highest organizational buy-in about knowledge management was found among professionals and knowledge workers. Senior management was next to recognize the importance of knowledge management, with middle management lagging behind. Team-based cultures were found to be much more amenable to knowledge sharing. Dow Chemical, Kaiser Permanente, and Texas Instruments were good examples of applying the team-based approach. Additionally, providing financial rewards to promote and reward sharing behaviors may not be as effective as embedding knowledge development and transfer in the employee's professional and career development systems.

Another finding from the study was that both push and pull approaches to knowledge management should be done together, vs. using only one of these techniques alone. Push approaches desire to capture knowledge in central repositories and then push it out into the organization. Pull approaches expect people to seek the knowledge they need.

Wiig (1997), one of the founding fathers of the knowledge management movement, provides some concluding views on where knowledge management is heading in the following years:

> "We cannot expect that the knowledge society will be the last societal or management evolution. We do not yet know what the next turn of events will result in, but we can expect that the knowledge management focus — even after it has been assimilated into the normal daily work tasks — will be pushed into the background by new priorities and approaches. It is reasonable to expect that such changes will take place in 25 to 30 years. When that happens, the need to manage knowledge cannot be expected to disappear. Instead, we will most likely find that management of knowledge processes and knowledge assets, along with management of intellectual, financial, physical, and other assets will have become routine with well-developed tools, practices, and monitoring approaches."

References

Davenport, T. and L. Prusak (1998), *Working Knowledge*, Harvard Business School Press, Cambridge, MA.

Halal, W., J. Artz, J. Liebowitz, and M. Kull (1998), *Organizational Intelligence: Toward a Conceptual Framework for the Cognitive Functioning of Complex Organizations*, George Washington University, Prepared for the National Science Foundation.

Mahe, S. and C. Rieu (1998), A Pull Approach to Knowledge Management, Proceedings of the 2nd International Conference on Practical Aspects of Knowledge Management, Basel, Switzerland, October 29–30.

Mentzas, G. and D. Apostolou (1998), Managing Corporate Knowledge: A Comparative Analysis of Experience in Consulting Firms, Proceedings of the 2nd International Conference on Practical Aspects of Knowledge Management, Basel, Switzerland, October 29–30.

Reimus, B. (1996), Knowledge Sharing Within Management Consulting Firms, Kennedy Publications, New Hampshire.

Wiig, K. (1997), Knowledge Management: Where Did It Come From and Where Will It Go?, *Expert Systems With Applications Journal*, Elsevier/Pergamon Press, Vol. 13.

Questionnaire

Role and skills for Knowledge Management

From TFPL, Ltd.

International Research Project

Questionnaire

Please complete the following if different from above:

Organisation: [ ]
Name: [ ]
Position: [ ]
Address: [ ]

Telephone: [ ]
Fax: [ ]
Email: [ ]

Is there anyone else in your organisation who we should contact? Yes ☐ No ☐

...if *Yes*, please give contact details:

Name: [ ]
Position: [ ]
Telephone: [ ]

Would you like to receive an executive summary of the final report? Yes ☐ No ☐

Part 1   Organisation

1.1   Type of organisation

☐ Academic
☐ Professional partnership
☐ Public sector
☐ Private sector *(please specify)*

☐ Other *(please specify):*

1.2   What is the geographic nature of your organisation?

☐ National
☐ Multinational
☐ Global

1.3   How is your organisation structured for corporate governance?

☐ Single / national global Head Quarters
☐ Series of geographical Head Quarters
☐ Series of specialist Head Quarters
☐ No Head Quarters (please describe below)

1.4   What is the approximate total number of employees within your organisation? *(please specify below)*

1.5   Where are the other major geographic locations of your organisation and approximate number of employees?

Role and Skills for Knowledge Management—Questionnaire

| Name of Country

 (*please specify*) | Main Activity
(eg: Head Quarters; Manufacturing; R&D, Regional Office) | Approx no. of employees at location
(*please specify*) |
|---|---|---|
| | | |
| | | |
| | | |
| | | |
| | | |
| | | |
| | | |
| | | |
| | | |
| | | |
| | | |

1.6 Approximately what percentage of your organisation's workforce spend one day (or more) a week travelling or working away from their office?

　　　　[   ] %

Part 2 Knowledge Management activities and development

We recognise that many organisations have programmes or initiatives not labelled KM but which could be regarded as KM activities. If this is the case, please indicate in question 2.2 then answer the following KM questions.

2.1 Do you have a corporate wide KM programme? Yes ☐ No ☐ Planned ☐

2.1.1 If *Yes*,
What is this programme called? *(please specify below)*

2.1.2 If *No*, do you have another corporate programme with similar objectives?
(eg: learning organisation, business process redesign, inclusive approach) Yes ☐ No ☐

2.1.2.1 If you answered YES to 2.1.2 above, what are these other programmes with similar objectives based on? *(tick all that apply)*
☐ Business units or practice areas
☐ Communities of interest
☐ Functional structure
☐ Location
☐ Other *(please specify)*

2.2 Are your KM activities geographically localised? Yes ☐ No ☐

2.3 Which of the following stages of KM development best matches your organisation?
☐ Stage 1 - Exploration of potential & interest
☐ Stage 2 - Planning strategy
☐ Stage 3 - Implementation
☐ Stage 4 - Established
☐ Stage 5 - Revisiting

2.4 Is knowledge management even across divisions and locations within your organisation, or have some units leading KM initiatives?
☐ Even development
☐ Uneven development

2.5 What are the main objectives of your Knowledge Management programme?
(please tick all that apply)
☐ Culture change

Role and Skills for Knowledge Management—Questionnaire **77**

- ☐ Improved creativity and innovation
- ☐ Improved market knowledge
- ☐ Improved speed of product/service development
- ☐ Improved utilisation of intellectual assets
- ☐ Process improvement
- ☐ Other *(please specify)*

2.6 What are the main features of your Knowledge Management programme? *(tick all that apply)*
- ☐ Identification and exchange of best practice
- ☐ Improved information sharing
- ☐ Learning organisation
- ☐ Management and exploitation of intellectual assets
- ☐ Skills training and development
- ☐ Other *(please specify)*

2.7 Who in the organisation has **championed** KM developments? *(please tick all that apply)*

At Board / Partner level:
- ☐ CEO
- *Directors or Partners responsible for:*
- ☐ Human Resources
- ☐ Information Technology
- ☐ Marketing
- ☐ Business Development
- ☐ Research & Development
- ☐ Specific Business Unit / Practice
- ☐ Other *(please specify)*

At Senior Management level:
- *Managers of:*
- ☐ Human Resources
- ☐ Information Technology
- ☐ Marketing
- ☐ Business Development
- ☐ Research & Development
- ☐ Specific Business Unit / Practice
- ☐ Other *(please specify)*

2.8 Does your organisation have: *(please tick all that apply)*

☐ A Chief Knowledge Officer ⇒ ☐ In a global role
⇒ ☐ In a national role
⇒ ☐ In a corporate role
⇒ ☐ In a divisional or practice role

☐ A Chief Knowledge Team
☐ A core implementation team
☐ A Central Knowledge Centre or Hub
☐ A number /network of Knowledge Centres / Hubs
☐ Formal knowledge networks

2.8.1 Who has responsibility for the implementation of KM activities?

Job title: [   ]

2.8.2 To whom does this person report? [   ]
 to?

2.9 How many people are directly involved in KM activities?

Globally: [   ]

Nationally: [   ]

2.10 Does your organisation have: *(please tick all that apply)*

☐ Formal Communities of Interest
☐ Informal Communities of Interest
☐ Formal Communities of Practice
☐ Informal Communities of Practice

2.11 Which functions in the organisation do you consider require intensive use of information or knowledge?

☐ Administrative   ☐ Marketing
☐ Competitive Intelligence   ☐ Production
☐ Human Resources   ☐ Research & Development
☐ Information Technology

2.12 Does your organisation have formal mechanisms for identifying skills and competency requirements? **Yes No**
☐ ☐

2.13 What mechanisms does the organisation use to develop knowledge management related skills?
☐ Appraisal

- [ ] Training programmes
- [ ] Mentoring
- [ ] Other *(please specify)*

2.14 Does your organisation have recognised procedures for "publishing" information internally, either in hardcopy or electronic format? **Yes** **No**

Part 3   KM roles and skills

3.1 The model below represents one organisational approach to structuring knowledge management roles, with the Corporate or Global CKO and planning team as the driver of the knowledge management programme. Other models may be driven from a geographic regional division.

Please illustrate the structure of your own organisation's KM roles, by modifying this model, or drawing your own in the space provided:

Knowledge Management Roles Structure

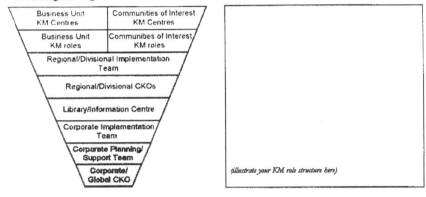

(illustrate your KM role structure here)

3.2 The table below contains generic KM roles with brief descriptions of their level and scope. Please complete the following:

	Nature of role: FT Full-time NR New role for an existing post FR Future role	Salary band: 1 £70,000+ 2 £40,000-69,999 3 £25,000-39,999 4 <£25,000	Post title (please specify)	Background (please specify)	Qualification (please specify)
1.	**Chief Knowledge Officer (CKO)** *Senior Executive responsible for KM leadership*				
1.1	Corporate / Global CKO *Corporate Board or Partner level*				
1.2	National CKO *National Board or Partner level*				
1.3	Business / Functional Unit CKO *Senior Management level*				
	Nature of role: FT Full-time NR New role	Salary band: 1 £70,000+ 2 £40,000-69,999	Post title	Background	Qualification

Role and Skills for Knowledge Management—Questionnaire

		for an existing post FR Future role	3 £25,000-39,999 4 <£25,000	(please specify)	(please specify)	(please specify)
2.	**Chief Knowledge Team (CKT)** *Senior management staff with responsibility for KM strategy and planning.* **Please specify additional roles:**					
2.1	HR process development issues					
2.2	IT / IS infrastructure issues					
2.3	Information content issues					
2.4	Organisational development issues					
3.	**Knowledge Management Team** *Senior / Middle management staff with responsibility for implementation of KM activities*					
3.1	Knowledge Director *Oversee development of processes, infrastructure and information resources*					
3.2	KM Project Manager *Responsible for facilitating the execution of agreed KM activities*					
3.3	KM Process Manager *Support and monitoring for KM processes*					
4.	**Knowledge Centre(s) / Hub(s)** *Teams responsible for the acquisition, dissemination, access, etc. of internal and external information and knowledge*					
4.1	Knowledge Centre / Hub Manager *Responsible for facilitating access to internal and external information and knowledge sources*					
4.2	Knowledge Co-ordinator *Responsible for acquiring internal and external information and knowledge in specific areas*					

	Nature of role: FT Full-time NR New role for an existing post FR Future role	Salary band: 1 £70,000+ 2 £40,000-69,999 3 £25,000-39,999 4 <£25,000	Post title (please specify)	Background (please specify)	Qualification (please specify)
4.3 Intranet developer(s)					

	Responsible for the design, development and navigation of the corporate Intranet					
4.4	Content editor Responsible for the structure, classification or codifying of internal information					
4.5	Database / Notes designer					
4.6	Journalist Responsible for debriefing and synthesising project experience					
4.7	Researcher					
4.8	Analyst					
4.9	KM help desk					
4.10	KM training					
5.	**Knowledge network roles** *(including communities of interest, practices and competencies)*					
5.1	Knowledge Leader Responsible for facilitating KM activities within the network, community and network building					
5.2	Knowledge Manager Facilitates access to internal/ external information and knowledge within a specific network					
5.3	Knowledge Navigator Responsible for knowing the information and knowledge resources of the specific network					
		Nature of role: PT Full-time NR New role for an existing post FR Future role	**Salary band:** 1 £70,000+ 2 £40,000-69,999 3 £25,000-39,999 4 <£25,000	**Post title** *(please specify)*	**Background** *(please specify)*	**Qualification** *(please specify)*
5.4	Knowledge Synthesiser Responsible for abstracting and synthesising current or new industry, subject, and client knowledge					

Role and Skills for Knowledge Management—Questionnaire

5.5	Knowledge Sponsor *Responsible for resourcing and authoring knowledge assets*					
5.6	Knowledge Facilitator *Responsible for sharing and collecting knowledge*					
5.7	Knowledge Owner *Responsible for specific knowledge assets*					
6.	**Business / Functional Units**					
6.1	Knowledge Leader *Responsible for facilitating KM activities within the network*					
6.2	Knowledge Manager *Responsible for facilitating access to internal and external information and knowledge within a specific network*					
6.3	Knowledge Collector / Navigator *Responsible for knowing the information and knowledge resources for the specific network*					
6.4	Knowledge Synthesiser *Responsible for abstracting and synthesising current or new industry, subject, and client knowledge*					
6.5	Knowledge Sponsor *Responsible for resourcing and authoring knowledge assets*					
6.6	Knowledge Facilitator *Responsible for sharing and collecting knowledge*					
6.7	Knowledge Owner *Responsible for specific knowledge assets*					

3.3 Would you or one of your colleagues be prepared to participate further in the project by providing further information on the skills required by the people in KM roles?

Type of Interview **Yes** **No**
Telephone ☐ ☐
Face-to-face ☐ ☐

3.4 Are there any further comments you would like to make:

Appendix A

The Intelligent Agent-Based Knowledge Management System for Supporting Multimedia Systems Design on The Web

Seung Ik Baek, Jay Liebowitz, Srinivas Prasad and Mary Granger
Management Science Department
School of Business and Public Management
The George Washington University

Marshall Lewis
Electronic Learning Facilitators, Inc.

AIS '97 Technology Demonstrations

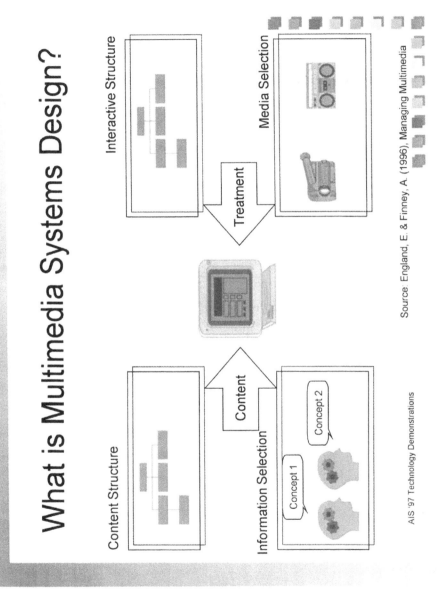

What is Knowledge Management (KM) in Multimedia Systems Design?

Data	Information	Knowledge	Knowledge Management
Text, audio, video, numbers, graphics, etc.	Text, audio, video, numbers, and graphics that are closely related to a specific topic.	1) What information should be contained (Content Knowledge). 2) How the information should be presented (Treatment Knowledge).	A method for systematically and actively managing and leveraging design ideas and decisions among team members while developing storyboards.

Why is KM Difficult in Multimedia Systems Design?

- Successful multimedia systems design requires a collaborative effort among many designers with a variety of backgrounds (ex., users, project managers, instructional designers, content experts, and media specialists).
- Successful multimedia systems design demands the balance between individual task accomplishment and group project achievement.

AIS '97 Technology Demonstrations

How to Support KM in Multimedia Systems Design?

- A KM system should speed the evolution of design knowledge by aiding the expression, transmission, and evaluation of individual ideas.

- A KM system should be suited for distributed group member use; it should be easily accessible by team members at any time and from any place.

- A KM system should enable active collaborative work among team members, rather than providing a passive information space.

This research will develop a KM system that can help design team members create, exchange, and share their storyboards on the Web.

AIS '97 Technology Demonstrations

Appendix A

Intelligent Agents

- Reactive
- **_Goal-Oriented_**
- **_Personality_**
- **_Self-Starting_**
- **_Cooperative_**

- **_Autonomous_**
- Flexible
- Adaptive
- Mobile

Intelligent agents are software programs that perform a given set of tasks on behalf of a user or other agents without the direct human intervention, and in so doing, employ some knowledge of user's goals.

Source: Wooldridge, M.L. & Jennings, N.R. (1996), Intelligent Agents

AIS '97 Technology Demonstrations

Knowledge Management Agents (1 of 2)

KM Agent	Possible Support Features
User Agent	• Remember all user activities. • Dynamically organize a user's agenda. • Validate user input.
Knowledge Agent	• Index design knowledge. • Detect inconsistency; generate recommendation. • Save, retrieve, and update design knowledge.

Knowledge Management Agents
(2 of 2)

KM Agent	Possible Support Features
Knowledge Manager	• Monitor all changes that occur in a knowledge repository and forward them to the user agent. • Reformulate queries based on an ontology (MM Design Knowledge). • Dynamically retrieve annotations, and generate hyper-links for them. • Manage versions of storyboards in a knowledge repository.

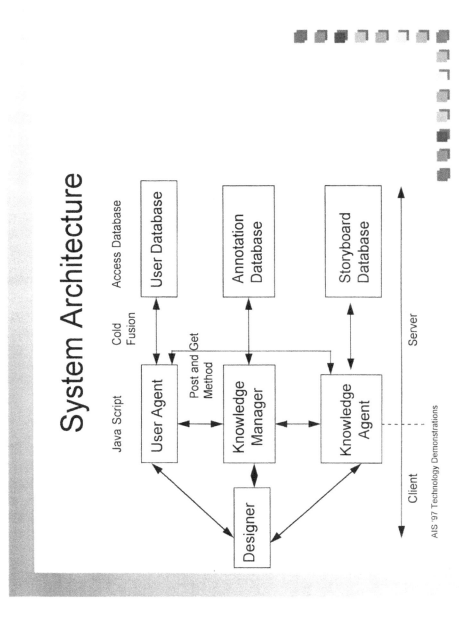

Appendix A 95

Agent Architecture

```
<CFQUERY NAME="GET_STORYBOARD" DATASOURCE="SB">
    SELECT * FROM STORYBOARD
    WHERE STORYBOARD_ID=#URL_NO#
</CFQUERY>

<HTML><BODY>

<SCRIPT LANGUAGE="JAVASCRIPT">

FUNCTION A ()
{    DOC.DOCUMENT.WRITE('<B> …… </B>'); ……}

IF (ID == "10000") {A;}
ELSE { B;}

</SCRIPT> ……..
<FORM NAME="CREATE_SB" ACTION="KMGR.CFM" METHOD="POST">
<INPUT TYPE="HIDDEN" NAME="ID" VALUE="#NEW_ID">

</FORM>
```

AIS '97 Technology Demonstrations

welcome to *elf*

The Intelligent Knowledge Management System
for Multimedia Systems Design

Your Browser Name: Netscape
Your Browser Version: 4.01 [en] (WinNT; I)

To use the system properly, you need 3.0 or higher version of Netscape.
Please check the version of your browser.

If you do not have User Name and Password, please a system administrator at seung@gwis2.circ.gwu.edu

Please Enter Your User Name and Password.

User Name: SEUNG
Password: ****

OK   Reset

Appendix A

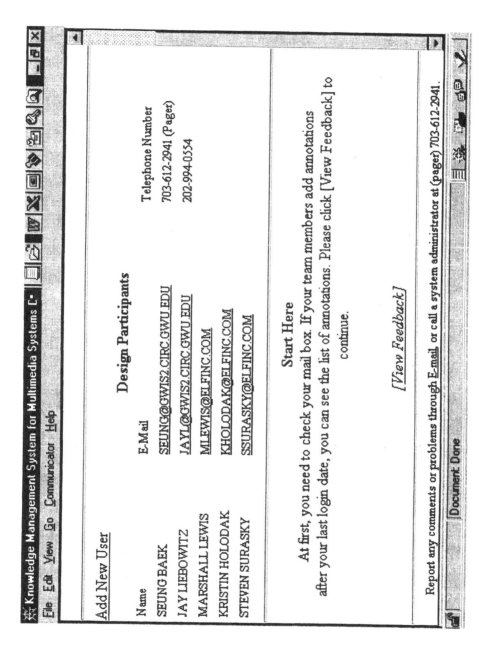

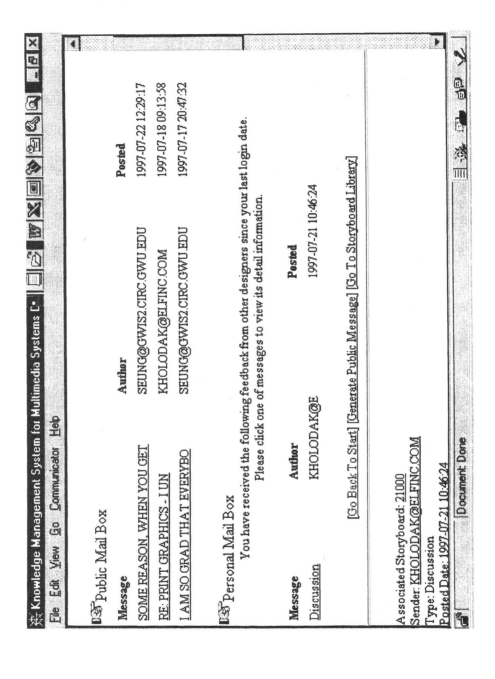

Appendix A

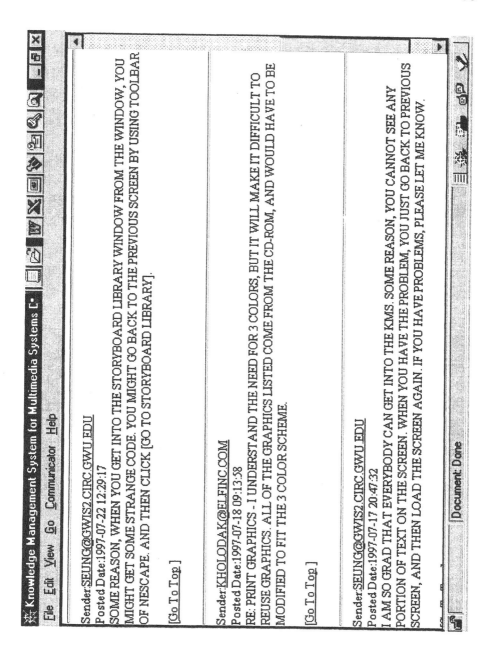

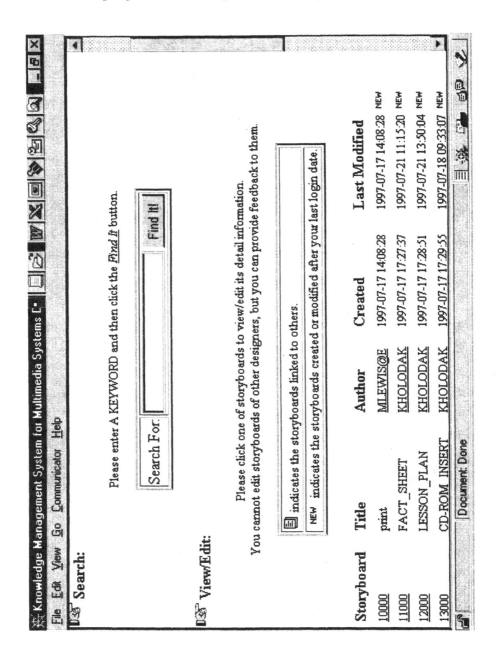

Appendix A

Knowledge Management System for Multimedia Systems L
File Edit View Go Communicator Help

Please describe the storyboard below: (Please do not insert new lines between sentenses)

INTRODUCTORY TEXT, FACT SHEET, CONTACT INFORMATION

[Save New Description]

Interactivity

[Show Help]

Hot Spot	Action	Result	Script/Go To		Last Modified
G13100	ONLOAD	GRAPHIC	THE MAIN INTERFACE W	0	18-Jul-97
T21000A	ONLOAD	TEXT	THE NOT AGAINS: DISAS	1 NEW	18-Jul-97
Add New Hot Spot					18-Jul-97
Add New Hot Spot				0	21-Jul-97
Edit Hot Spot				0	21-Jul-97
Edit Hot Spot					

Document: Done

Interactivity [Show Help]

Hot Spot	Action	Result	Script/Go To	Last Modified
G13100	ONLOAD	GRAPHIC	THE MAIN INTERFACE W	18-Jul-97
T21000A	ONLOAD	TEXT	THE NOT AGAINS: DISAS	18-Jul-97
L21000C	CLICK	GOTO	24000	21-Jul-97
L21000B	CLICK	GOTO	23000	21-Jul-97

General Comments:

To view detail information, please click one of the underlined message types.
To add your response for a message, please click the message and then click [Respond].

Type	Topic	Author	Posted	No. of Feedback
CONTENT NOTE	G21000B EXAMPLE	KHOLODA	18-Jul-97	0
CONTENT NOTE	HYPERTEXT LINKING	KHOLODA	21-Jul-97	1 NEW

[Go Top] [Add General Comments]

Appendix A

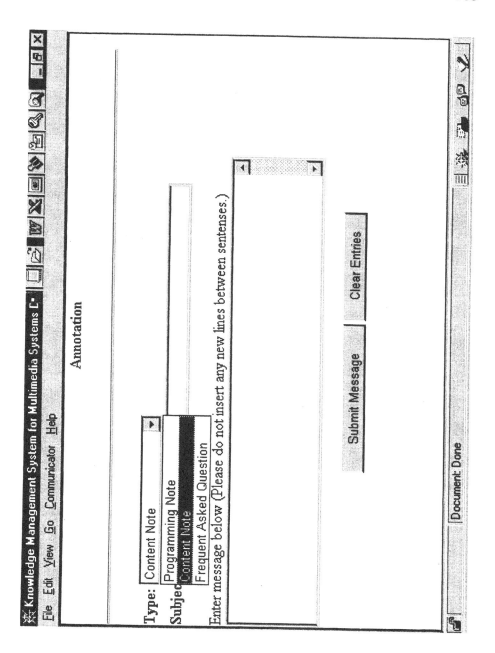

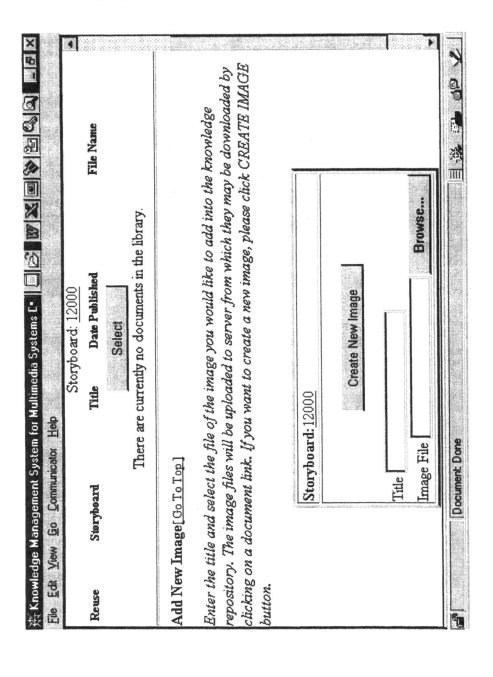

Appendix A

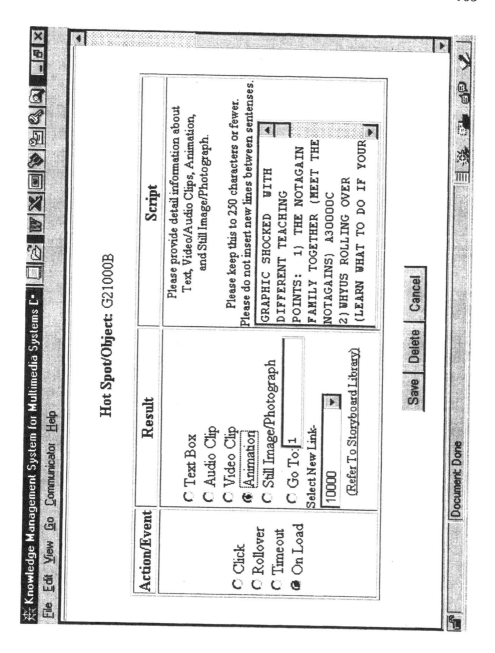

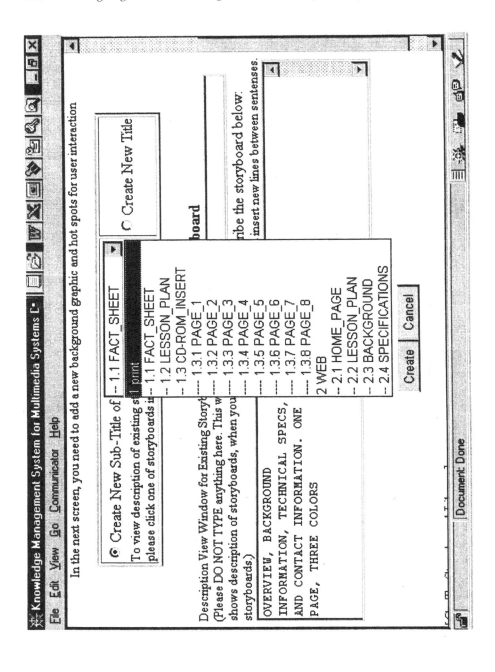

Storyboard Library

Similar storyboards already exist - Please Check Them

The following storyboards have LESSON in their title, description, or objects/hot spots

Storyboard	Comment	Title	Author	Creation Date
22000	Check Title and Description	LESSON_PLAN	KHOLODAK@ELFINC.COM	18-Jul-97
12000	Check Title	LESSON_PLAN	KHOLODAK@ELFINC.COM	17-Jul-97
13000	Check Description	CD-ROM_INSERT	KHOLODAK@ELFINC.COM	17-Jul-97
13400	Check Description	PAGE_4	KHOLODAK@ELFINC.COM	18-Jul-97
13500	Check Description	PAGE_5	KHOLODAK@ELFINC.COM	18-Jul-97

New Storyboard: 30000
Title: LESSON
Description:
LESSON PLAN

[Add New] [Cancel]

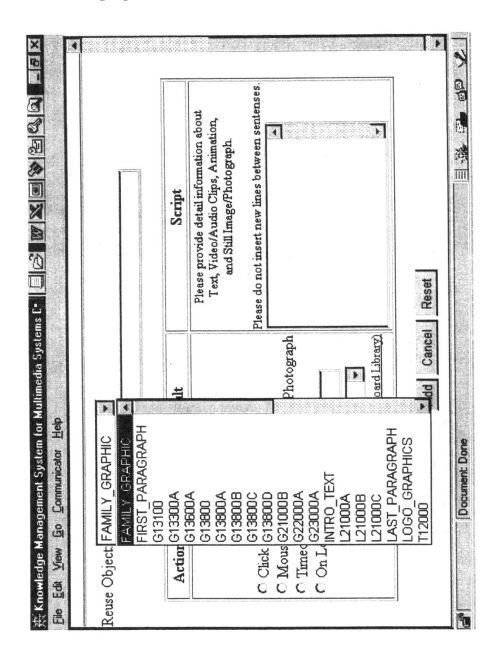

Appendix A

Storyboard	Link From	Link To
10000: print		
11000: FACT_SHEET		
12000: LESSON_PLAN		
13000: CD-ROM_INSERT		
13100: PAGE_1		
13200: PAGE_2		
13300: PAGE_3		
13400: PAGE_4		
13500: PAGE_5		
13600: PAGE_6		
13700: PAGE_7		
13800: PAGE_8		
20000: WEB		
21000: HOME_PAGE	21000 (L21000A)	L21000A => 22000
	21000 (L21000B)	L21000B => 23000
		L21000C => 24000
22000: LESSON_PLAN		
23000: BACKGROUND		

Storyboard No: 13200   **Storyboard Title:** PAGE_2
Creator: KHOLODAK@ELFINC.COM

Description:
START THE FACT SHEET

Base Graphic File:

Interactivity

Action	Hot Spot	Result	Script/Go To
ONLOAD	T13200B	TEXT	THIS CD-ROM IS PART OF A JOINT EFFORT BETWEEN THE AMERICAN RED CROSS AND PHILLIPS PETROLEUM. THE BOMBING OF THE FEDERAL BUILDING IN OKLAHOMA CITY PROVED THAT PEOPLE CAN RESPOND WELL IN EMERGENCIES, BUT THERE ARE ALWAYS NEW WAYS TO HELP PEOPLE EVEN MORE
ONLOAD	T13200A	TEXT	THE NOTAGAINS: DISASTER PREPAREDNESS AND

Future Work

- Integrate more design heuristics.
- Integrate case-based reasoning.
- Support multiple design projects.
- Support a whole design process.

Appendix B*

Appendix B

AUDIT OF KNOWLEDGE MANAGEMENT PRACTICES FOR INNOVATION

This questionnaire provides the basis for an audit of the Knowledge Management Practices (KMPs) that play an important role in shaping and making available the stocks of knowledge which are used for innovation in a firm. Each KMP should be assessed in terms of its contribution to the creation and use of these evolving knowledge bases in the organisation or unit being evaluated.

- **In the left-hand column**: Rate your degree of satisfaction with the current performance of this KMP in contributing to the organisation's evolving knowledge bases. If the KMP is not implemented currently, leave this column blank.
- **In the middle column:** Identify the level of importance you attach to this KMP in supporting the organisation's evolving knowledge bases.
- **In the right-hand column**: Tick this box if some significant action might need to be taken in relation to this KMP. The document Actions for KMP Improvements offers guidelines on following up key requirements identified from the completed questionnaire.

Group A: R&D Management Activities

This first group of KMPs focuses on the formal and informal R&D management activities which also play a significant role in the accumulation and management of project experience, results and expertise.

	Satisfaction Low – High	Importance Low – High	Action
A1 Technical reports from R&D projects are archived in a form that makes them easily accessible to relevant personnel (eg in a library; within a structured database).	1 2 3 4 5	1 2 3 4 5	
A2 Controlled access to project-specific R&D knowledge is offered:			
A2a by topic-based query systems (eg using cross-project libraries)			
A2b to company staff outside the local project/department			
A2c to selected people/organisations outside the company			
A2d soon after confirmed findings have been first recorded			
A3 An assessment is carried out at the end of each project to identify factors affecting success or failure which go beyond purely technical information (eg failures in information flows; contextual information about suppliers, customers, collaborators, equipment, etc).			
A4 Strategic cross-project reviews explicitly utilise and contribute towards the evolving stocks of corporate knowledge (eg as source of intelligence on internal capabilities, external technologies, markets and competitors).			
A5 The outputs of formal R&D management control activities (eg milestone reviews; en-of-project assessments) are also actively utilised as sources of technical, project and commercial information.			
A6 Intermediate data and results, gathered before producing formal project reports, are made available to relevant personnel (eg through online data feeds to other projects).			
A7 Intermediate and final interpretations of project findings are distributed to relevant personnel other than local project and departmental management (eg on an ad hoc basis using targeted distribution list to contact individuals who are likely to be interested in a specific finding).			
A8 Quality monitoring or other auditing activities are routinely used to provide systematic feedback on best practice (eg on maintenance of test equipment).			

*Developed by the University of Manchester, U.K.

	Satisfaction	Importance	Action
	Low High	Low High	
A9 Training and guidance is provided to technical staff in the communications skills needed to present and disseminate data, ideas, interpretations and findings to a variety of audiences (eg creating and using intranet 'Research Web' sites; writing non-technical 'White Paper' summaries of technological innovations; creating 'value-added intelligence' reports on competitors).	1 2 3 4 5	1 2 3 4 5	
A10 Results from technical modelling and design activities (eg diagrams; physical prototypes; software simulations) are used to improve communications between staff from different disciplines and functions, as well as for their specific purpose within a project.			
A11 Outcomes (eg new techniques or ideas) that are additional or peripheral to specified project outputs: A11a are explicitly identified from project experience A11b have their potential explored A11c are disseminated to relevant personnel (eg through Feedback notes)			
A12 The physical location of R&D personnel and support staff is arranged in groupings designed to promote knowledge creation and transfer (eg clustering together personnel who span project, departmental or other organisational boundaries; using a centrally-placed library as an informal communication node).			
A13 Management provides practical support (eg computer networks; physical spaces; funding) to foster the development of ad-hoc arrangements that also contribute towards knowledge creation and transfer (eg creating 'formal informality' through internal special-interest groups; 'coffee-time' discussions; topic-based electronic bulletin boards).			
A14 Individuals or groups wishing to foster a novel approach or technology have access to formal company procedures which allow them to argue the business case for allocating fresh resources to that innovation.			
A15. Specific personnel are allocated ring-fenced time for defined cross-boundary gatekeeping activities			
A16 'Gatekeepers' in key positions gather, communicate and contextualise knowledge across disciplinary and organisational boundaries (eg technology, customer and competitor stewards).			
A17 The work of gatekeepers inside and outside R&D is explicitly co-ordinated.			

Group B: Mapping Knowledge Relationships Across Boundaries

R&D activities often span disciplinary, organisational and company boundaries. The KMPs in this Group facilitate the co-ordination of internal and external R&D capabilities, inter-firm relationships and market requirements. Many use 'capability maps' (textual and/or graphical) which identify the locations of formal knowledge bases, people-embodied skills and supporting equipment and services.

	Satisfaction	Importance	Action
	Low High	Low High	
B1 Regular analyses of the 'capabilities for innovation' (knowledge, people, equipment/services) in the unit are made independently from decision making about specific projects.	1 2 3 4 5	1 2 3 4 5	
B2 These analyses are used to create and maintain clear capability maps in the areas essential to the organisation's R&D success (eg technical capabilities, staff skills, projects, products, users, market sectors).			
B3 Mapping information is disseminated widely (eg steward-run intranet Web sites; topic-centred electronic and hardcopy newsletters).			
B4 The existence of these maps and how they can be used is known by all relevant staff.			
B5 Capability maps are updated continuously to represent long-term accumulations of knowledge (eg through a Technology Audit procedure).			
B6 Relationships and alliances between competitors are routinely monitored to update relevant capability maps.			
B7 External maps incorporate capabilities in universities and other non-commercial sources (eg through scanning the public science base).			
B8 The company's capabilities are regularly compared to those of its competitors (eg using capability maps to guide technology sourcing strategy).			
B9 Capability maps are used regularly to rate current potential against strategic R&D and business aspirations.			
B10 Managers make strategic decisions using 'landscape' studies that include long-term 'roadmaps' of the relationships between technological developments and the requirements of products, customers and markets.			
B11 R&D staff gain and maintain direct knowledge of customer and market requirements (eg by providing external consultancy or online user support).			
B12 R&D staff participate in studies of the characteristics, behaviour and needs of users and customers in order to identify promising areas of technology and product development.			
B13 Selected R&D staff are allocated the time and resources to work closely with external standards-setting and regulatory bodies.			
B14 Internal peer review is used to judge: B14a internal developments and new ideas in science and technology			
B14b potential external sources of science and technology			
B15 Specific groups and activities are arranged with the prime aim of bringing together people from different disciplines, projects and organisational units (eg Expert Panels with a variety of specialists; Design Reviews covering multiple projects; regular videoconferences spanning organisational boundaries).			

Group C: R&D Human Resource Management

These KMPs relate to the motivation, reward and appraisal of R&D personnel. They can play a key role in creating and using the corporate knowledge bases by providing incentives to share knowledge, develop inter-disciplinary expertise and undertake cross-boundary working.

	Satisfaction Low High	Importance Low High	Action
C1 Personnel involved in matrix management or dual-reporting processes are equally rewarded for achieving both specialist departmental goals and project targets.	1 2 3 4 5	1 2 3 4 5	
C2 The secondment of R&D personnel (eg to product development or user support units) is actively managed as a feature of both career progression and knowledge transfer.			
C3 Staff are rewarded for good performance in disseminating the results of their research within the company and, where appropriate, externally.			
C4 Managers undertaking staff appraisals are required to give positive weighting to successful performance outside an individual's main discipline.			
C5 Individuals' CVs are maintained and archived in a form that is easily accessible to relevant personnel (eg for use by a project manager in assembling a new team or a new employee looking for expert advice).			
C6 Special training is provided in synthesising knowledge from a variety of multi-disciplinary sources.			
C7 Personnel who are explicitly identified as experts in a specific field who can act as internal 'consultants' outside of project-specific assignments:			
C7a are allocated time and resources to conduct their consulting tasks			
C7b keep records of all consultancy enquiries they receive			
C7c use these records to trigger future research (eg using Feedback Notes)			

Group D: Intellectual Property Management

Legal requirements, such as the 'Priority of First Proof' in US Patent Law, demand a number of KMPs to handle Intellectual Property (IP) issues. The following KMPs utilise and extend these activities to contribute towards the accumulation and management of knowledge.

	Satisfaction Low High	Importance Low High	Action
D1 IP specialists automatically and regularly update R&D staff with information on patents and patent applications relevant to their areas of technical specialisation (eg a targeted 'Patent Watch' service; up-to-date information on IP positions of competitors and potential partners or acquisitions).	1 2 3 4 5	1 2 3 4 5	
D2 IP issues are addressed explicitly at all milestones in project progression (eg 'Prior Art' IP evaluation during the Feasibility stage).			
D3 Summaries and analyses of the patent situation in particular fields are provided pro-actively by IP specialists to assist strategic and project decision making.			
D4 IP specialists provide guidance and training on generic IP issues and techniques to R&D project staff.			
D5 Ad hoc IP searches in specialist areas are carried out quickly and accurately (eg through search software available to all staff).			
D6 Routine and ad hoc IP searches in generic or cross-project technologies are performed quickly and accurately.			
D7 IP specialists are routinely informed about the emergence of potential novelty.			
D8 Lab Books and R&D Notebooks (to satisfy IP requirements) are written and archived in a form that makes their content easily accessible to relevant personnel.			

Group E: R&D Information Management

The way information is recorded, stored and structured for retrieval and communication is a critical element in supporting most KMPs in Groups A to D. In recent years, a variety of interconnected and rapid innovations in IT and Communication Technologies (ICTs) have offered opportunities to transform these information management practices. The following KMPs highlight the role of ICT-based information management as a trigger to creating or changing KMPs.

		Satisfaction		Importance		Action
		Low	High	Low	High	
E1	A corporate information strategy to build and disseminate the evolving corporate knowledge bases is continuously monitored and updated at all appropriate levels in the company.	1 2	3 4 5	1 2 3	4 5	
E2	Access to specific ICT-based services is controlled according to clearly stated and widely known policies.					
E3	The whole company uses a common IT infrastructure (eg standardised operating system and applications; seamless transfer between different systems).					
E4	ICT systems supporting collaborative working (eg groupware) are easily available and usable for all those involved with R&D projects.					
E5	Librarians and information scientists collaborate routinely with IT specialists in developing and implementing the information management strategy.					
E6	Resource budgeting for information management includes library and information science requirements, as well as IT specialists and systems.					
E7	Information management activities take account of the value of paper-based and lower-tech systems (eg the telephone and fax), as well as more sophisticated IT capabilities.					
E8	Direct input from R&D staff contributes to the development of information management activities (eg by online feedback to suggest opportunities for new systems and applications; regular meetings between R&D and information management staff).					
E9	R&D staff are able to experiment with new ICT-supported KMPs (eg using email or videoconferencing to build ad hoc virtual teams or special interest groups).					
E11	Software simulations and other IT demonstrators are used to encourage cross-boundary working by making tangible the opportunities that can be gained by improving communications between staff from different disciplines and functions.					
E12	Information management specialists routinely provide 'information navigation' advice in the early stages of a project.					
E13	The following ICT-based services are used by the organisation to build, analyse and disseminate the organisations' evolving knowledge bases:					
	E13a internal unstructured email					
	E13b internal structured email (eg 'threaded' email discussion groups')					
	E13c a company intranet (eg internal Research Web)					
	E13d simple remote access to internal information systems for all R&D staff when outside the company's physical boundaries					
	E13e an 'extranet' allowing controlled access to specific areas of the company intranet by external people (eg commercial and academic collaborators)					
	E13f universal access for R&D staff to the Internet					
	E13g tailored support for navigating the World Wide Web and other Internet sources (eg distributing files to relevant personnel containing 'bookmarks' to sites of interest to them)					

E13h	easy-to-use mechanisms for targeting information dissemination (eg tailorable email dissemination lists)			
E13i	access to external databases			
E13j	a document management system for archiving and retrieving knowledge			
E13k	videoconferencing			
E13l	online access to scientific journals			
E13m	Patent Trend Analysis software			

4. <u>IMPROVING KMPS FOR INNOVATION</u>

Results obtained from the KMP for Innovation questionnaire can be interpreted in ways which deliver practical improvements only if the analysis is based on sound knowledge of the specific context in which improvements are being sought. It is therefore not possible to provide general prescriptive advice on how these results should be treated. The 'Actions for KMP Improvements' form on the following pages is offered as a general aid to self-auditing which companies can adapt to their own requirements. As with the questionnaire, it has been designed so that it can be used on its own, although it should preferably be used in conjunction with the whole of this document.

Appendix B

ACTIONS FOR KMP IMPROVEMENTS

The *Audit of KMPs for Innovation* questionnaire identified those practices where improvements are needed. By analysing the satisfaction and importance ratings of these, the KMPs where improvements are most likely to deliver the major benefits can be identified. These key KMPs (usually nor more than ten), provide an initial for developing action plans to follow up the audit. This could be done for the organisation as a whole and/or for individual units.

The kinds of actions to be taken will probably be unique to each environment. The form overleaf provides a termplate format for specifying action plans.

- **KMP No. and Description** relate to the KMPs as identified in the Audit Questionnaire.
- **Start New KMP Activity** highlights where a totally new KMP or knowledge management activity within a current one should be started.
- **Alter Existing KMP** specifies changes to be made to an existing KMP. Typically these will involve increasing or decreasing the degree of formality, the levels of access, and/or the overall scope of the current KMP.

An example of how the form could be completed is provided below.

KMP No.	Description	Start New KMP Activity (initiate full new KMP, or activity within existing KMP)	Alter Existing KMP (e.g. increase or decrease formality/access/scope)
A3	End of Project Assessment	1. Place all EOPAs on the Research Intranet. 2. Alert Quality Team to any project experience which requires changes to company standards.	1. Re-emphasise importance of EOPAs in training and project management guidelines 2. Redefine formal EOPA processes to clarify feedback procedures outside the project. 3. Extend scope of EOPAs to include non-technical feedback (e.g. people management).
B11	Customer/Market experience for R&D staff	Require all staff providing external consultancy to keep a 'day book' on queries answered.	Relax constraints on when R&D staff can be in direct liaison with customers and users.
E2	ICT access control		Revise policy to take account of Research Web
E13c	Research Intranet	Start project to build internal 'Research Web'	

ACTIONS FOR KMP IMPROVEMENTS

KMP No.	Description	Start New KMP Activity (initiate full new KMP, or activity within existing KMP)	Alter Existing KMP (e.g. increase or decrease formality/access/scope)

Appendix C

Appendix C

**Knowledge Management:
Strategies for the Learning Organisation
Screening Survey**

In the context of this study, we consider "Knowledge Management" to include the strategies, tactics, and support mechanisms for the creation, identification, collection, and sharing of knowledge and practices and applying the best knowledge within the organisation. The purpose of Knowledge Management is to improve your organisation's effectiveness by leveraging the knowledge you have and need to use to compete.

Depending on your business and strategy, important knowledge can range from the intellectual assets that underlie products and services; knowledge about your customers and markets; the identification and transfer of "best practices"; and individual expertise.

We believe knowledge is a fundamental factor in the effectiveness of modern organisations, and want to understand how they address it.

Name: _____ Phone: ( ) _____
Job Title: _____ Fax: ( ) _____
Organisation: _____
Full Address: _____
City/State/Zip: _____
E-mail (if any): _____

Primary Industrial Sector:
- ❏ Aerospace/Defense
- ❏ Automotive
- ❏ Banks/Financial
- ❏ Food & Beverage
- ❏ Services/Hotel/Retail
- ❏ Software
- ❏ Chemicals/Petroleum
- ❏ Consulting/Accounting
- ❏ Healthcare/Pharmaceutical
- ❏ Telecommunications/Utilities
- ❏ Government
- ❏ Other: _____
- ❏ Consumer Goods
- ❏ Transportation
- ❏ Insurance
- ❏ Computers/Electronics
- ❏ Manufacturing

If your organisation is unfamiliar with the term "Knowledge Management," are there other descriptions/definitions given to knowledge initiatives internally? ❏Yes ❏No
If so, what are they? (please tick the boxes)

- ❏ Intellectual Capital
- ❏ Patent Management
- ❏ Learning Organisation
- ❏ Other, please specify: _____

Section I: General Information

1. **This survey is being completed for** (*check one*):
   - ❏ Total Organisation
   - ❏ Business Unit/Plant Site only

Please complete the following questions (2-3) for the total organisation or business unit/plant site specified in question 1. If you wish, you may complete a copy of this survey for the total organisation with additional copies for one or several business units or plant sites.

2. **What are the total number of employees and annual revenues (converted in 000$) for the most recent fiscal year?**

	# of employees	# of white collar employees that are Knowledge Workers	Annual Gross Revenues in $000.00
Total Organisation			
Business Unit/Plant Site			

3. **In what geographical area does your organisation primarily operate** (*check all that apply*)?

   - ❏ US only
   - ❏ Primarily in North America
   - ❏ Western Europe only
   - ❏ All of the above: we operate globally
   - ❏ Asia
   - ❏ Africa
   - ❏ Western and Eastern Europe
   - ❏ South America

Appendix C

Section II: Strategy, Approaches and Process

4. Do you know/measure the value of your organisation's intellectual capital?
   - ❏ Yes   ❏ No

5. What are the most important Knowledge-carriers in your organisation?
   - ❏ People ❏ Paper ❏ Magnetic Media ❏ Processes ❏ Products and Services ❏ Other, please specify_____

6. Do your organisation's overall strategic goals include knowledge management explicitly?
   - ❏ Yes   ❏ No

   If yes, are there people assigned to knowledge management (implicitly or explicitly) ? Please specify the functions (currently involved) and explain briefly:
   _____
   _____
   _____

7. How long have you had a knowledge management initiative?
   - ❏ less than one year
   - ❏ 2-4 years
   - ❏ Other, please specify_____
   - ❏ 1-2 years
   - ❏ 4 years or more

8. What is currently the strategic emphasis within your organisation ?
   - ❏ Operational Excellence and Cost-Leadership (= focusing on providing a well-balanced products/services-mix, with an optimal combination of price, quality and comfort).
   - ❏ Product and Technological Leadership (= focusing on continuous improvement and continuous renewal of products/services, based on "state-of-the-art"-technology .
   - ❏ Customer Intimacy (= focusing on providing "total" solutions for a well-selected group of customers)
   - ❏ Other, please specify:_____

9. Please rate the following "Knowledge-Management" objectives in the context of your Business-Strategy ? *(please rate between : 1= highest importance and 6 = lowest importance)*
   - ____ Facilitation of the "re-use" and consolidation of Knowledge about operations
   - ____ Standardisation of existing Knowledge in the form of procedures/protocols
   - ____ Combination of Customer-Knowledge and internal know-how
   - ____ Acquisition of New Knowledge from external sources .
   - ____ Generation of New Knowledge inside the organisation
   - ____ Transforming individual (people's) knowledge into collective knowledge
   - ____ Other, please specify:_____

10. What approaches do you use to improve your knowledge assets and operations?

a) **Sharing and combination of knowledge:**

❏ External or Internal Benchmarking	Is this approach:	❏ Organisational-wide	❏ Business Unit Specific
❏ Communities of Practice (expert-groups)	Is this approach:	❏ Organisational-wide	❏ Business Unit Specific
❏ Cross-Functional Teams	Is this approach:	❏ Organisational-wide	❏ Business Unit Specific
❏ Intranets (including Groupware)	Is this approach:	❏ Organisational-wide	❏ Business Unit Specific
❏ Training and Education (including Corporate Learning Centers)	Is this approach:	❏ Organisational-wide	❏ Business Unit Specific
❏ Documentation and Newsletters	Is this approach:	❏ Organisational-wide	❏ Business Unit Specific
❏ Other, please specify:_____	Is this approach:	❏ Organisational-wide	❏ Business Unit Specific

b) **Creation and Refinement of Knowledge:**

❏ Lessons learned analysis (i.e. project reviews)	Is this approach:	❏ Organisational-wide	❏ Business Unit Specific
❏ Learning spaces/labs	Is this approach:	❏ Organisational-wide	❏ Business Unit Specific
❏ Research and development center	Is this approach:	❏ Organisational-wide	❏ Business Unit Specific
❏ Explicit Learning Strategy	Is this approach:	❏ Organisational-wide	❏ Business Unit Specific
❏ Other, please specify:_____	Is this approach:	❏ Organisational-wide	❏ Business Unit Specific

c) **Storing of Knowledge:**

❏ Storage of customer knowledge	Is this approach:	❏ Organisational-wide	❏ Business Unit Specific
❏ Best practice inventories	Is this approach:	❏ Organisational-wide	❏ Business Unit Specific
❏ Lessons learned inventories	Is this approach:	❏ Organisational-wide	❏ Business Unit Specific
❏ Manuals and handbooks	Is this approach:	❏ Organisational-wide	❏ Business Unit Specific
❏ Yellow pages of Expertise/Knowledge	Is this approach:	❏ Organisational-wide	❏ Business Unit Specific
❏ Other, please specify:_____	Is this approach:	❏ Organisational-wide	❏ Business Unit Specific

Appendix C

11. Are there specific knowledge management-related examples, activities, practices, functions, or capabilities that you consider to represent your best knowledge management efforts?

   A1. Briefly describe the first practice area and its stage of implementation

   _____
   _____
   _____

   a) Is this practice area: ❑ Organisational Wide or   ❑ Business Unit Specific

   b) How long have you been working on this practice area?:_____

   c) Who initiated this practice area?:_____

   d) How and when did implementation begin?:
   _____
   _____
   _____

   e) What skill sets and expertise is needed for this practice area?:
   _____
   _____

   f) Why does management continue to support this practice area ?:
   _____
   _____
   _____

   g) What are the bottom-line impacts of this practice area ?
   _____
   _____
   _____

B1. Briefly describe a second example or practice area and stage of implementation

a) Is this practice area: ❑ Organisational Wide or ❑ Business Unit Specific

b) How long have you been working on this practice area?:_____

c) Who initiated this practice area?:_____

d) How and when did implementation begin?:
_____
_____
_____

e) What skill sets and expertise is needed for this practice area ?:

f) Why does management continue to support this practice area ?

g) What are the bottom-line impacts of this initiative?

Section III: Culture

12. Which aspects of your organisational culture seem to support effective knowledge management?

13. Which aspects of your culture seem to be barriers to effective knowledge management?

14. To what extent has there been organisational buy-in and acceptance about knowledge management at the following levels?

Mgmt Senior mgmt	☐ None	☐ A little	☐ Some	☐ A lot	☐ Total
Middle mgmt	☐ None	☐ A little	☐ Some	☐ A lot	☐ Total
Supervisory mgmt	☐ None	☐ A little	☐ Some	☐ A lot	☐ Total
Staff Professionals/Know-ledge workers	☐ None	☐ A little	☐ Some	☐ A lot	☐ Total
Operational & craft workers	☐ None	☐ A little	☐ Some	☐ A lot	☐ Total

15a. Which cultural factors did you explicitly take into consideration when implementing your knowledge management strategy, what spects of culture needed to be changed and how was this managed?

_____
_____
_____
_____

15b. Do you have specific training programme in place to support Knowledge Management?

_____
_____

16. Has your organisation taken steps to motivate and reward people and/or teams supportive of effective knowledge management?

    ❏ Yes   ❏ No

17. Please list or describe briefly any incentives/reward systems that support knowledge management.

_____
_____
_____
_____

18. What aspects of your culture changed as a result of the implemented Knowledge Management process?

_____
_____
_____

19. What insights or "lessons learnt" have you experienced thus far into culture changes?

_____
_____
_____

Section IV: Technology

20. **Do you use IT as an enabler to (check all that apply):**
    - ☐ Investigate, assess, safeguard important knowledge
    - ☐ Use the best knowledge to do the job well
    - ☐ Learn and innovate to do the job better
    - ☐ Reengineer the workplace and the production system
    - ☐ To create new products and services
    - ☐ Other, please specify_____

21. **Do you have a "formal" WebMaster function?**
    - ☐ Yes   ☐ No

22. **Which of the following technologies do you use to support knowledge management ambitions?**

    ☐ **INTRANET** (Internal Internet based upon WWW browsers or groupware)
      - ☐ E-mail
      - ☐ Video Conferencing
      - ☐ Yellow pages of expertise (knowledge maps)
      - ☐ Discussion Forums
      - ☐ Shared documents/products
      - ☐ Training and Education
      - ☐ Gathering and publication of lessons learnt/best practices

    ☐ **INTERNET FUNCTIONS:**
      - ☐ Knowledge searching on WWW
      - ☐ Business Intelligence
      - ☐ Knowledge exchange with suppliers
      - ☐ Knowledge exchange with customers

    ☐ **KNOWLEDGE AND DATABASES**
      - ☐ Knowledge based systems (including Experts systems and case-bases)
      - ☐ Competency information systems
      - ☐ Best Practice/Lessons learn database (If yes, is the Best Practice database centralised or decentralised?)_____
      - ☐ Customer information systems

    ☐ **DATA MINING AND KNOWLEDGE DISCOVER TECHNOLOGIES:**
      - ☐ Extracting customer knowledge from databases for marketing purposes
      - ☐ Extracting knowledge from process data to improve operations.
      - ☐ Simulation, interactive multi media.

130 *Building Organizational Intelligence: A Knowledge Management Primer*

23. Are these knowledge and databases accessible through the Intranet?

    ❑ Yes   ❑ No

    If yes, which one(s) ?

    _____
    _____
    _____

24. Are you planning to make these knowledge and databases accessible through the Intranet?

    ❑ Yes   ❑ No

    If yes, which one(s)?

    _____
    _____
    _____
    _____

25. Which technologies do you *plan to use* to support knowledge management?

    _____
    _____
    _____
    _____

Section V: Business Outcomes

26. Do you capture, measure and track the value of your organisation's knowledge?
    ❏ Yes   ❏ No
    If yes, please explain:
    _____
    _____
    _____
    _____

27. Do you measure/track the (New) Knowledge-generation within your organisation?
    ❏ Yes   ❏ No
    If Yes , please explain:
    _____
    _____
    _____

28. Do you measure/track knowledge sharing within your organisation?
    ❏ Yes   ❏ No
    If yes, please explain:
    _____
    _____

29. Do you have a measurement system that shows how knowledge management affects bottom line?
    ❏ Yes   ❏ No

30. Do you measure the value added of KM and the cost of?
    ❏ Yes   ❏ No

If yes, how do you define and measure the "value-added" of Knowledge Management ?

_____
_____
_____

31. **Where do you see the desired Business-benefits, for your organisation, through the systematic Management of Knowledge and Intellectual Capital ?** (please rate the following "benefits" in order of importance - 1= very important , 6 = not so important for your organisation)

   ____ Lower Operating Costs

   ____ Increased Revenue / Sales

   ____ Higher Sales-Effectiveness

   ____ Improving (Quality) products and services

   ____ Innovation (New products and services)

   ____ Cycle-time reduction

   ____ Time-to-market improvements

   ____ Improved (strategic/tactical) decision making

   ____ Other, please specify:_____

32. **Do you have any defined measures of effectiveness (focusing on *outcomes*) of your KM strategy?**
    ❑ Yes   ❑ No

32a. **If yes, please indicate which of the following outcomes your organisation has realized as a result of its KM practices(check all that apply):**

❑ **Increased innovation** (increased % of sales from new products; reduced cycle time to product release)
   Measured by: ❑ In-depth Analysis, ❑ Anecdotal, Evaluations/Survey, ❑ Other_____

❑ **Business Growth** (successful new business ventures, growth in sales to existing customers, increased cash flow/market share)
   Measured by: ❑ In-depth Analysis. ❑ Anecdotal. ❑ Evaluations/Survey, ❑ Other_____

❑ **Practice & Process improvement** (decreased cycle time, overall improvement in quality, increased productivity, more reliable information)
   Measured by: ❑ In-depth Analysis, ❑ Anecdotal, ❑ Evaluations/Survey, ❑ Other_____

Appendix C 133

- ❑ **Increased Customer Satisfaction**
  Measured by: ❑ In-depth Analysis, ❑ Anecdotal, ❑ Evaluations/Survey, ❑ Other_____

- ❑ **Enhanced employee capability and organisational learning** (rapid/innovative problem-solving, more accurate/timely problem solving, improved mobility, improved capabilities of individuals/teams)
  Measured by: ❑ In-depth Analysis, ❑ Anecdotal, ❑ Evaluations/Survey, ❑ Other_____

- ❑ **Other:** *(please specify)* _____
  Measured by: ❑ In-depth Analysis, ❑ Anecdotal, ❑ Evaluations/Survey, ❑ Other_____

33. **Is knowledge management integrated with business processes or product processes?**

    ❑ Business   ❑ Product

34. **Are there any other points you would like to make or areas that we have not mentioned that would help us understand your situation with regard to knowledge management?**

    _____
    _____
    _____
    _____

35. **Would you consider hosting a "site-visit" on your knowledge management experiences, within the context of this project?**

    ❑ Yes   ❑ No

Appendix D

Appendix D

Valuating Knowledge Assets (Human Capital)
KNOWLEDGE SURVEY

1. If we define "knowledge assets" to be associated with only "human capital", the following factors probably contribute to human capital growth. For each factor, indicate in a few lines as to how the organization contributes to each factor (i.e., how each factor relates to the organization):

A. TRAINING AND EDUCATION:

1. Formal Training of Employees

   _____
   _____
   _____

2. Formal Education of Employees (i.e., degrees)

   _____
   _____
   _____

3. Mentoring and On-the-Job Training

   _____
   _____
   _____

B. SKILLS:

1. Research Skills

   _____
   _____
   _____

2. Entre- and Intra- preneurship Skills

   _____
   _____
   _____

3. Retention Rates of Employees

   _____
   _____
   _____

C. OUTSIDE PRESSURES & ENVIRONMENTAL IMPACTS:

1. Industry Competition

2. Half-life of Information in Industry

3. Demand and Supply of Those in the Field

D. INTERNAL & ORGANIZATIONAL CULTURE:

1. R&D Expenditures of the Organization

2. Formalized Knowledge Transfer Systems (e.g., lessons learned databases or best practices guidelines institutionalized within the organization)

3. Informal Knowledge Transfer Systems (e.g., speaking often with top management, secretaries and assistant to top management, attending organizational events, the "grapevine")

4. Interaction with Customers and Users

5. Physical Environment and Ambiance (e.g., nice office, reasonable resources, etc.)

Appendix D **137**

6. Internal Environment Within the Organization (e.g., reasonableness of demands by management placed on the employees, etc.)

_____
_____
_____

7. Short Term (2-4 years) and Long Term (5 years or more) Goals/Prospects, from the Employee's perspective, of the Organization's Viability and Growth

_____
_____
_____

E. PSYCHOLOGICAL IMPACTS:

1. Morale (attitude, benefits, compensation, conferences, travel, vacation time, etc.) Of Employees

_____
_____
_____

2. Creativity and Ingenuity of Employees

_____
_____
_____

3. Employee Stimulation and Motivation

_____
_____
_____

(CONTINUED ON NEXT PAGE)

II. Please rank the factors from 1 (most important) to 19 (least important) in terms of their importance, from your perspective, towards contributing to human capital growth:

___formal training of employees
___formal education of employees
___mentoring and on-the-job training
___research skills
___entre- and intra-preneurship skills
___retention rates
___industry competition
___half-life of information in industry
___demand and supply of those in the field
___R&D expenditures of the organization
___formalized knowledge transfer systems
___informal knowledge transfer systems
___interaction with customers and users
___physical environment and ambiance
___internal environment within the organization
___short term and long term goals
___employee morale
___employee creativity and ingenuity
___employee stimulation and motivation

III. For each of the factors above, please rate how well the organization is satisfying the top 5 factors mentioned above (please use Excellent, Good, Fair, Poor).

Factor 1: _____
Factor 2: _____
Factor 3: _____
Factor 4: _____
Factor 5: _____

Thanks for completing this questionnaire. Kindly send it to Professor Jay Liebowitz, Dept. Of Information Systems, UMBC, 1000 Hilltop Circle, Baltimore, Maryland 21250.

Index

A

Aggregation Model for OI assessment, 11
American Management Systems, 12–13, 57
Analytic Hierarchy Process, 39
Andersen Consulting, 56–57
Audits
    benefits of, 65–66
    factors to consider, 65
    objectives of, 64
    value of, 63–64

C

Capturing and storing knowledge
    inadequate system, case study illustrating, 23–29
    interviewing, 20
    knowledge vs. data/information, 19
    learning by doing, 21
    mechanisms for, 21
    observation and simulation, 20–21
    protocol analysis, 20
    questionnaires and surveys, 20
    storage approaches, 29–30
Case-based reasoning, 2, 34
Chief Knowledge Officer, 3, 4, 22, 58
Combining knowledge, 38–39
Community of Practice, 7–8, 57
Competitive Intelligence, 15
Corporate culture, *see also* Learning culture
    acceptance of knowledge management, 50–51
    development of, 49–50
    learning organization concept and, 35, 51–52
    negative factors influencing, 53
    positive factors influencing, 52
    use of incentives, 53
Cross-Level Model for OI assessment, 11

D

Data mining, 3
Distributed Model for OI assessment, 11–12
Distributing knowledge, 39–40

E

ELAWS, 3
Engineering Book of Knowledge (EBOK), 53–54
Executive learning, 35–36
Expert Choice, 39
Explicit knowledge, 14, 15
Explicit-to-Explicit Knowledge Transfer, 33
Explicit-to-Tacit Knowledge Creation, 33

H

Human capital, 1, 25

I

Intellectual capital, 7
Intellectual capital and organizational intelligence, 13
Intelligent systems, 2–3
Internet, 22
Interviewing for capturing knowledge, 20

K

KM, *see* Knowledge management
Knowledge attic approach, 28–30
Knowledge creation
    social theory elements, 16
    techniques for, 15–16
Knowledge management (KM)
    audits, 63–66
    capturing and storing knowledge, 19–21
    combining knowledge, 38–39
    company examples, 56–57, 69
    concerns for using, 4–5
    corporate culture and acceptance of, 50–51
    definition, 1, 3
    distributing knowledge, 39–40
    effective use of, 70
    elements of competencies, 12–13
    factors affecting, 2
    intelligent systems discipline applied, 2–3
    knowledge attic approach, 28–30
    knowledge creation elements, 33–34
    knowledge sharing environment, 1
    knowledge vs. data/information, 19–21
    learning incorporated into, 32
    organizational change and, 53
    organizational intelligence, relation to, 70
    reasons for implementing, 14, 64
    responsibility for, 59
    status of roles, 58
    storage approaches, 29–30
    strategic need for, 22
    strategies for, 55–56
    tacit/explicit knowledge, 14–16, 33
    transferring knowledge, 37–38
    transforming information, process of, 6
    trends in, 58
    using existing approaches, 56
Knowledge map, 33–34
Knowledge organization, 29, 67–68
Knowledge publisher, 30
Knowledge pump, 30
Knowledge sponge, 30

L

Learning by doing for capturing knowledge, 21
Learning culture, *see also* Corporate culture
    characteristics of, 42
    contribution to knowledge environment, 43
    contribution to organizational memory, 43
    elements of, 41–42
    need for, case study illustrating, 44–46
Learning organization, *see also* Organizational learning
    corporate culture and, 51–52
    elements of, 31–32
    executive learning, stages of, 35–36
    need for, case study illustrating, 44–46
    problematic learning, 33
    supportive culture needed, 35, 51–52
    types of learning, 32
Lessons learned, case study using, 45–46

M

Multiple cooperating expert systems, 34

O

Observation and simulation for capturing knowledge, 20–21
Organizational intelligence (OI)
    definitions of, 3, 5–6, 11
    intellectual capital and, 7, 13
    key factors and functions, 68–69
    knowledge functions involved, 6
    knowledge management and, 70
    organizational learning role in, 8–10
    renewal process, 7–8
Organizational learning, *see also* Learning organization
    elements of, 9
    individual learning transformed into, 34
    obstacles to, 9–10
    requirements for, 55
    role of, 8–10
    social capital contributions to, 9
Organizational memory, 43

P

Protocol analysis for capturing knowledge, 20

Q

Questionnaires and surveys for capturing knowledge, 20

R

Repositories
    importance of, 38
    need for, case study illustrating, 45

S

Senior management commitment, 4, 35–36
Sharing culture, 9
Social capital, 8–9
Storage approaches for knowledge, 29–30, *see also* Capturing and storing knowledge
Structural capital, 1

T

Tacit knowledge, 14–15
Tacit-to-Explicit Knowledge Creation, 33
Tacit-to-Tacit Knowledge Creation, 33
Transferring knowledge
    importance of knowledge repository, 38
    transfer function, 37

V

Value creation, strategic need for, 12, 13

W

Warrant, 38
Wisdom, 22
World Bank, 57

WisdomBuilder

**KNOWLEDGE ISN'T POWER
WISDOM IS!**

Desktop Edition

WISDOM BUILDER IS THE FIRST IN A FAMILY OF KNOWLEDGE MANAGEMENT TOOLS THAT PROVIDE THE MARKET'S ONLY FULLY INTEGRATED SOLUTION TO THE INFORMATION MANAGEMENT AND ANALYSIS DILEMMA. WISDOM BUILDER SOLVES THE INFORMATION OVERLOAD PROBLEM BY REDUCING THE TIME AND COST OF EXTRACTING INFORMATION AND OTHER RESEARCH KNOWLEDGE FROM UNORGANIZED REPOSITORIES OF HETEROGENEOUS DATA.

PRODUCT HIGHLIGHTS

- AUTOMATES DATA COLLECTION
- CONVERTS DATA TO KNOWLEDGE
- VISUALIZATION OF INFORMATION RELATIONSHIPS
- PROTECTS CORPORATE KNOWLEDGE
- COLLABORATIVE USER ENVIRONMENT

WISDOM BUILDER RUNS UNDER WINDOWS 95, 98, AND NT.

SUGGESTED RETAIL PRICE $2,495

www.wisdombuilder.com

WisdomBuilder LLC

GOLD RUSH!

45-Day Evaluation Edition

NOTHING BUT NUGGETS!

GOLD RUSH! IS THE "*SLOWEST* SEARCH ENGINE OF THE NET" DESIGNED FOR SERIOUS INTELLIGENCE AND INFORMATION GATHERING FROM BOTH THE INTERNET AND FOLDERS AVAILABLE ON THE USER'S PC OR NETWORK. GOLD RUSH! CAN BEST BE DESCRIBED AS AN "AI-BASED META SEARCH ENGINE" AS IT COMBINES THE ATTRIBUTES OF ARTIFICIAL INTELLIGENCE (NATURAL LANGUAGE PROCESSING) AND INVOKES MULTIPLE SEARCH ENGINES.

PRODUCT HIGHLIGHTS
- NATURAL LANGUAGE INPUT
- REAL-TIME WEB SCANNING
- SHOW ME MORE!
- SEARCH HISTORY
- EXPORTING SEARCH RESULTS

GOLD RUSH! RUNS UNDER WINDOWS 95, 98, AND NT, REQUIRES INTERNET EXPLORER 4.01 OR LATER AND AN INTERNET CONNECTION FOR WEB BASED SEARCHES.

SUGGESTED RETAIL PRICE $79.95

www.wisdombuilder.com

WisdomBuilder LLC